P9-DXD-575

As Dr. Reese makes clear ... we must look to ourselves to safeguard our remarkable planet for future generations.

—Dr. James Hansen, director of NASA Goddard Institute for Space Studies, author of *Storms of my Grandchildren*

As chilling as a howl in a moonlit wood, *The Insatiable Bark Beetle* is a desperate plea for sense to prevail.

—Steve Payne, editor, Australian Broadcasting Corporation's *Organic Gardener*

In this lucid and information-rich book, Dr. Reese Halter tells the story of an ancient relationship gone awry, perhaps the most dramatic example to date of how climate change is disrupting and unbalancing the Earth's ecosystems.

—David Perry, professor emeritus, Oregon State University, co-author of *Forest Ecosystems*

The Insatiable Bark Beetle is a well-written, systematic examination of the growing challenges we humans face by hiding behind the intellectual wall of informed denial and social irresponsibility with respect to global warming.

—Chris Maser, zoologist, co-author of *Economics and Ecology: United for a Sustainable World*

Can it be true that a handful of fertile soil contains more micro-organisms than the total number of humans who have ever lived? Can beetles, birds and trees be linked in a way that can transform our world? This small book has huge implications for our global future.

—Robyn Williams, award-winning Australian science journalist and broadcaster, author of *True Story Waiting to Happen*

Dr. Reese Halter has done it again. Not satisfied with bringing global attention to what honeybees have been trying to tell us, in his most recent book *The Insatiable Bark Beetle*, he is acting as microphone for the tiny bark beetles. Read Dr. Reese's new book, in which he beautifully and compassionately tells these stories, and then do something to help fight climate change.

—Doug Larson, professor emeritus, University of Guelph, author of *Storyteller Guitar*

The passion (and despair) of the author pervades the book. There may be the temptation by some to mistake the passion for ideology. This would be a mistake. The book is backed by quality science. Politicians and policy makers should read this book. I thoroughly recommend it.

—Roger Sands, professor emeritus, University of Canterbury, author of *Forestry in a Global Context*

The Insatiable Bark Beetle is a warning that the unintended consequences of climate change are already with us and are reaching deep into our forests. The balance of nature between plants and insects he describes is fascinating, and an important reminder of the interconnectedness of life on Earth. A great read!

—Robert Teskey, distinguished research professor, Warnell School of Forestry & Natural Resources, University of Georgia

Dr. Reese has a knack to simplify complex situations and describe in a poetic, simple and inviting manner the enchanting areas of the world he has studied or visited. To all busy and curious human beings I recommend reading a few pages of Dr. Reese's stories on a regular basis. They will give you a positive outlook on life.

—Aldo Bensadoun, founder and CEO of the ALDO Group

The
INSATIABLE
Bark
Beetle

Dr. Reese Halter

RMB

Victoria Vancouver Calgary

Rocky Mountain Books
www.rmbooks.com

Library and Archives Canada Cataloguing in Publication

Halter, Reese
  The insatiable bark beetle / Reese Halter.

Includes bibliographical references.
Issued also in electronic format. (ISBN 978-1-926855-67-7)
ISBN 978-1-926855-66-0

  1. Bark beetles. 2. Bark beetles—Effect of global warming on.
3. Trees—Diseases and pests—Environmental aspects. 4. Trees—Diseases and pests—Economic aspects. 5. Insect populations—Climatic factors. I. Title.

SB945.B3H35 2011   634.9'6768   C2011-903305-4

Printed and bound in Canada

Rocky Mountain Books acknowledges the financial support for its publishing program from the Government of Canada through the Canada Book Fund (CBF) and the province of British Columbia through the British Columbia Arts Council and the Book Publishing Tax Credit.

  Canadian Heritage   Patrimoine canadien   Canada Council for the Arts   Conseil des Arts du Canada

 BRITISH COLUMBIA ARTS COUNCIL

The interior pages of this book have been produced on 100% post-consumer recycled paper, processed chlorine free and printed with vegetable-based dyes.

 FSC MIX Paper from responsible sources FSC® C016245

For my true love and soul mate – Lovey

All things are connected. Whatever befalls the Earth befalls the children of the Earth.

CHIEF SEATTLE
SUQWAMISH AND DUWAMISH

Contents

ith oth
a massive,
opulation

Introduction

About 10,000 years ago, 50 per cent of Earth's land surface was forested. Today a little over 25 per cent of the planet is covered with trees. Humans are destroying and polluting the biosphere at an extraordinary rate. In fact, each day we are spewing 82 million metric tons of greenhouse gases into our stratosphere. Yet, some of us have the temerity to claim no responsibility for our daily rapacious activities.

The first decade of the 21st century made its mark in history by being the warmest decade since 1880, when modern temperature measurement began. The second decade is already shaping up to distinguish itself along similar lines, as the global average temperature for 2010 was 14.63°C (58.3°F), which tied with 2005 as the hottest year over the past 130 years. During the nine decades between 1880 and 1970, the global average temperature increased by about 0.03°C (0.05°F) each decade, for a total increase of approximately 0.27°C (0.49°F). However, since 1970 the rate of increase in global average temperature has shot up to 0.13°C (0.23°F) per decade, for a total increase of around 0.5°C (0.9°F) in just over 40 years. Thus, over the past 13 decades, the average temperature on Earth has increased by a total of almost 0.8°C (1.4°F). While we may often consider this a relatively small number, it is a distressingly large increase in terms of climate, with widespread consequences for all life on Earth.

All biological systems are complex and very sensitive to temperature. Each system works in concert with others to produce exquisite ecosystem services that benefit not only humankind, but all life forms. For those of us who have worked for decades observing and documenting wild aquatic and terrestrial ecosystems, it is heartbreaking to witness the unravelling of these intricate biological systems. Scientists are now documenting drought- and heat-related forest stress (and death) on all forested continents. For the first time, it is possible to see frightening global patterns emerging.

Elevated temperatures, combined with other recent phenomena, are triggering a massive, worldwide increase in forest insect populations and diseases, in particular throughout western North America. Over the past 15 years, in excess of 24 million hectares (59 million acres) of mature forests have been killed by native bark beetles. These destructive insects – each about the size of a plump grain of rice – number in the hundreds of billions and have killed billions of mature pine, spruce and Douglas-fir trees. The beetles are opportunistic, and due to rising temperatures they are able to seek out water-stressed trees that have compromised defence systems and overcome them with their sheer numbers. Unfortunately, the frigid spring, fall and winter temperatures that would normally kill these bark beetles have not occurred for the past 15 years.

The most destructive of all the western North American bark beetle species – the mountain pine beetle (*Dendroctonus ponderosae*) – has already destroyed half of the commercial timber, the lodgepole pines, in British Columbia. Apart from the inconceivable volume of wood, enough to build several million homes, the beetles have taken a crucial terrestrial system that absorbs carbon dioxide (CO_2) – what's known in biological parlance as a "carbon sink" – and turned it into a "carbon source." When the dead trees decompose, they release CO_2 into the atmosphere. Over the next decade, the beetle-killed BC forests will emit 250 million metric tons of CO_2 – the equivalent of five years of car and light truck emissions in Canada. Imagine, this massive amount of excess CO_2 is from just one region of decomposing trees out of many around the globe.

All across western North America alone, bark beetles are devastating the forests like never before in modern times, perhaps in the entire history of conifers. Moreover, the beetles are advancing into regions of high-elevation forests, which typically rarely experience outbreaks. They are also venturing into the northern boreal forest, which has not evolved to defend itself against these voracious insects that kill trees synchronously, whole stands at a time. The future of our forests in western North America is precarious.

Extreme weather events are piling up around the globe. A German insurer, Munich Re, released a worldwide tally of 950 natural disasters in 2010, of which 90 per cent were weather-related. Total damage was estimated at in excess of $130-billion. But the costs do not stop there. Climate is also disrupting global food security, with every staple crop noticeably increased in price. China is suffering from the worst drought in decades and has stockpiled 60 million metric tons of grain, though even that is not enough for their 1.3 billion people for more than a year. World wheat markets have reacted by soaring over 80 per cent in the past 12 months. As we will see, rising temperatures are significantly affecting the yields and quality of coffee, tea, spices and many other commodities.

Human beings are exceptional problem solvers. Our innovations throughout the ages have been breathtaking. But despite our extraordinary abilities, there are issues that do not receive enough attention. For instance, a Stanford University project published in 2010 polled 1,372 scientists working in climate research and found that 97 to 98 per cent of them agree that humans are forcing Earth's climate by burning fossil fuels, releasing heat-trapping gases. Yet, on December 1, 2010, the first order of business in Washington,

DC, was to eliminate a Congressional committee on global warming. These types of developments are far too common and very discouraging.

Clearly we are locked into an energy model that is antiquated, subsidized and toxic to all life on Earth. We need citizens of all nations to work together toward change, as we are all interlinked. For instance, Australia is the largest supplier of coking coal to China, the country that has the largest carbon footprint on the globe. The USA is a close second, with the help of Canada's supplying 1.3 million barrels of crude oil and bitumen per day from the oilsands of northern Alberta. Canada's bitumen development is leaving a colossal carbon footprint, which has increased by 120 per cent since 1990. Of all industrial nations, Canada's footprint has increased the most during this time.

The forests of western North America and elsewhere around the planet, as well as life in the oceans, are vividly showing startling and tangible evidence of human-induced climate disruption. As an intelligent species, it is imperative that we understand what is happening. Change is imminent. We must adapt quickly or we – and life around us – will perish.

.peratur.
easons, the
g and ver

Forests

All wild forests are teeming with life. They are
biological treasure houses, easily equivalent in
riches to Fort Knox, the Museum of Modern
Art, the Louvre or the Prado. Much like a visiting a great museum, an individual can easily walk
through a forest without noticing the innumer-
able details that work in concert to generate
and sustain the majestic environment. Beyond
such a superficial glance, however, the layers of
interdependent life are bewitching. The composition of species and lifecycle processes in every
forest both creates and depends on the unique
combinations of all life in the area. An incredible
balance exists between the characteristics of all
components of a forest, from the tiniest insects,
to the amount of sunlight, to the various tree
species, to water – the lifeblood of the Earth.

While there are many classification systems

for forests, none are universally agreed upon. One simple way of dividing up the Earth's forests is by biomes. A biome is generally defined as a specific community of flora and fauna that occupies a large habitat. Earth contains seven naturally occurring biomes: tundra, taiga, temperate forest, tropical rain forest, desert, grassland and ocean. While an overview of all biomes is beyond the purview of this manifesto, both the taiga and the temperate forest biomes provide the setting for the story that I aim to illuminate: the story of the insatiable bark beetle.

Taiga, also known in parts as the northern boreal forest, makes up the largest uninterrupted, or contiguous, forested area on the surface of the Earth – the Earth's emerald crown. Stretching across Canada, through central Alaska, northern Asia, Russia, Europe, Scandinavia and northern Scotland, the taiga contains one-third of all the trees in the world. In sections, these forests are populated by the most widespread tree around the globe, the European aspen, as well as especially resilient species like the deciduous Siberian larch, which can withstand temperatures of −55°C (−67°F). These biological communities thrive in northern environments. While temperatures span a wide range throughout the seasons, the taiga biome is characterized by very long and very cold winters.

The temperate forest biome is located about halfway between the tropics and the poles, in both the southern and the northern hemispheres. In the southern hemisphere there are relatively small areas of temperate forest, in South America, Africa, New Zealand and Australia. In the northern hemisphere, temperate forests are primarily found throughout Europe, the eastern half of the United States, a small area of southern Ontario in Canada and through parts of Russia, China and Japan. These forests exist in relatively moderate climates that, throughout four distinct seasons, induce changes in forest life.

Temperate forests host a variety of trees that usually belong to three main groups: deciduous, coniferous and broad-leaved evergreens. Deciduous trees, like the broad-leaved species oak, chestnut, beech, maple and elm, are very efficient at trapping sunlight but not at tolerating freezing. As a result, they lose their leaves in the autumn, with a brilliant display of colours, and grow them back in the spring when the days get longer. Coniferous trees, also known as conifers or evergreens, develop cones to house their seeds and usually have needles for leaves. Cedars, firs and pines are all coniferous trees. Broad-leaved evergreens are primarily found in the southern hemisphere and include the eucalypts and tea trees.

The temperate climate zones of the Earth also host temperate rain forests. While these forests may contain broadleaf, coniferous or mixed species, the world's largest temperate rain forests, from northern California to Alaska, are predominantly conifers. In this region, the forests are home to the tallest trees on the Earth, including the coast redwoods, with heights that surpass 115 metres (377 feet), as well as cathedral-like Sitka spruce, western hemlocks, western red cedars and Douglas-firs. Along with these trees, millions of salmon, thousands of eagles and hundreds of wolves, black bears and giant grizzlies also thrive in the lush rain forests.

Every tree in the world participates in many biological processes, but one of the most notable is photosynthesis. In this process, tree leaves absorb carbon dioxide (CO_2) from the atmosphere, mix it with water, convert it into sugar using the sun's energy, and give off oxygen as a by-product. Carbon is stored in the wood during this process. In fact, trees are effectively the greatest CO_2 warehouses ever designed. For every metric ton of wood created, 1.5 metric tons of CO_2 is absorbed and 1 metric ton of oxygen is released. A team of scientists led by Oregon State University conclusively showed that Pacific Northwest old-growth forests are superlative at capturing and storing vast amounts of CO_2. Even after 200 years in existence, second-growth low-elevation forests could not compete with the magnificent storage capacity of the ancient forests.

Along with photosynthesis, trees are also involved in water purification. Nothing is comparable to the efficient filtration of fresh water that is completed by trillions of tree roots. The vapour-pressure deficit (suction) of the atmosphere pulls water molecules through the soil, up the trunk and into the leaves. Once in the leaves, specialized pores called stomata exchange water vapour into the air for CO_2 into the leaf. Big trees can pump over 450 litres (119 gallons) of moisture into the atmosphere every 24 hours. Leaves have ingenious architectural designs that minimize inundation of cells from precipitation and maximize water drip to the forest floor. Forest canopies intercept as much as 40 per cent of incoming annual precipitation, which decreases soil erosion.

Healthy soil is essential for all forest life. Tree roots actively secrete organic acids and carbohydrates to feed soil fungus, bacteria and complex soil micro-fauna. It is the most complete recycling process that we know of on the globe. There is no waste. Stated another way, there is zero unemployment. Every organism has a role to play: they all eat and they all reproduce. For example, the astonishing diversity of life on the forest floor in the Carmanah Valley on Vancouver Island enables Sitka spruce, the tallest trees in Canada, to reach over 95 metres (312 feet). In that forest alone there are 12 species of slugs – slimy herbivores that make up as much as 70 per cent of animal biomass. A square metre of the soil here can support 2,000 earthworms, 40,000 insects, 120,000 mites, 120,000,000 nematodes and many millions of protozoa and bacteria. They all "make a living" together: growing and working, they feed, digest, mate and die – all perpetuating the life of the forest.

In nature, diversity is essential. Coral reefs are among the most biologically diverse ecosystems, but arguably the most important place on earth for biodiversity is the Amazon jungle. Situated in South America, with portions in eight countries and one overseas territory – Brazil (~60 per cent), Peru (~13 per cent) and smaller areas in Colombia, Venezuela, Ecuador, Bolivia, Guyana, Suriname and France (French Guiana) – the Amazon is home to one in ten of the over 1.6 million known species on Earth. Furthermore, the Amazon River not only carries almost one-fifth of the world's flowing fresh water– generated in the surrounding rain forests and equal to that of the next ten largest rivers combined – it is also home to over 3,300 fish species, which is more than the entire Atlantic Ocean.

While tropical rain forests such as the Amazon occupy only 3 per cent of Earth's land-mass, they are the hotbed for evolution, bar none. Over 2,000 species of trees live in each hectare

(2.47 acres) of the Amazon. The forms of life are dazzling, with predator/prey interactions in constant arms races. For instance, tropical fishing bats use sonar to find fish and utilize a gaff-like toe to catch them, while plants have developed incredible chemical defences against fungi, bacteria, viruses, insects and animal grazers and browsers. By-products of all the diversity and interaction in tropical forests are countless and, more often than not, benefit humans to an extraordinary degree. These forests not only protect us from carbon dioxide and provide us with oxygen, but they are also the greatest pharmaceutical factories on Earth, regularly providing the opportunities for us to find new health products and potent medicines.

The existing Amazon climate is only "perfect" for year-round plant growth in a thin band within 5 degrees of the equator, known as the Intertropical Convergence Zone. The bright morning sunshine heats up the vegetation and evaporates water from it, resulting in the upward convection of wet air. As this air rises, it forms clouds and produces rain in the afternoon and evening. The climate is warm, wet and windless throughout the year. Half of the rain that falls on the great jungles in the Amazon River Basin returns to the atmosphere through the leaves of trees, to fall again as rain upon the Amazon forests. The daily cloud formations reflect a massive amount of the incoming solar radiation back into space; these clouds are of paramount importance for keeping the world cool. Something terribly wrong is happening in the Amazon right now, though. This Ferrari of jungles is breaking down with an alarming regularity and the effects are amplifying around the globe.

In 2010, British scientists made a significant breakthrough in the fight to save nature from senseless and rapacious destruction. Former banker Pavan Sukhdev and his team calculated a "price tag" for the worldwide network of environmental resources. Each biological system was assigned a dollar value for the services that are necessary to sustain humankind. Healthy rain forests, wetlands and coral reefs were valued at US$5-trillion per annum. Over the past 50 years, human activities have degraded at least 60 per cent of all ecosystem services globally. For instance, at current rates of deforestation in the

Amazon, it is estimated that 55 per cent of the region's rain forests could be gone by 2030. The extent of destruction is so extraordinary that new maps of the Earth are being drawn with boundaries determined not by natural systems but by the new, predominantly human-modified ecosystems. At an accelerating rate, humans continue to change natural systems by, for instance, plowing, grazing, fishing and hunting, timber extraction, river diversion, city expansion, water removal and the use of polluting fertilizers.

Losing our natural ecosystems and biodiversity is cause for serious "all hell's breaking loose" alarm. Our lives depend on nature and the natural world depends on the way we live our lives. Consider just a few examples of the benefits we receive from healthy, forested ecosystems:

- Fresh water is provided for billions of humans around the globe.
- Oxygen is generated, which enables all life on Earth to exist.
- Billions of tons of harmful, heat-trapping CO_2 – mostly created by burning fossil fuels – is

stored in the wood rather than staying in our atmosphere as pollution.

* Snow is retained and released slowly in the springtime, feeding agricultural systems.
* We discover powerful medicines that exist naturally in these environments, which help to fight cancer, coronary problems and other diseases.
* Billions of tourism dollars are generated, which help to sustain communities.
* Billions of dollars in non-timber forest products are harvested, such as greens for the floral industry, nutraceuticals, nuts, berries, mushrooms, honey, maple syrup, ginseng, fragrances, resin, rubber and the materials necessary for basket making and wood carving.

In order to flourish and provide these and other benefits, forests have specific requirements and optimal conditions that need to be fulfilled. For some, in western North America, Europe, Australia and elsewhere, fire is an integral part of their lifecycle. Many forests in North America, for example, have adapted to fire in various ways. Not all forests have had that opportunity though.

The wet tropical rain forests rarely experience fires, and about 3 per cent of western North American forests are considered non-fire regimes because fire is so rare that plant life has never been influenced by it.

Three types of forest fires are induced by lightning: crown fires, surface fires and ground fires. When treetops burn it is called a crown fire, which is deadly for most trees. However, some species are able to live through such trauma, such as the big-cone Douglas-fir and oaks in California that can regenerate leaves after being scorched. Other trees across the West have also developed strategies to take advantage of crown fires. Lodgepole pine cones in the Rocky Mountains and elsewhere – except for California, central Oregon and along the Pacific coast – open from the heat of a crown fire and release millions of seeds to colonize the burnt lands. Lodgepole pine forests are usually of the same age, because all of the trees started to grow at the same time following the fire. If fire is suppressed in lodgepole pine forests, however, then conditions are created for another agent of change to do the colonizing: the mountain pine beetle. The beetles enter the ecosystem and, as we will see, ultimately create new forests.

Unlike crown fires, surface fires burn from the top of the forest floor to about 3 metres (10 feet) above the ground. Certain trees, such as the western larch, Douglas-fir, ponderosa pine and the sequoias, have adapted to tolerate this type of fire. They hold their branches high in the crown, so that the flames of the surface cannot ignite the foliage. The mature trees also have bark at least 300 millimetres (12 inches) thick, which helps to protect the trunks of these trees from the heat of fast-moving surface fires. If you look closely at old Douglas-fir, ponderosa pine, western larch or sequoia trees, you'll notice they have blackened bark or fire scars indicating past occurrences of surface fires.

The ponderosa pine ecosystem in its southerly range may experience natural surface fires every 10 years or so. For about the past 100 years, though, a Smokey the Bear policy of stopping all fires across western North America has dramatically changed the structure of these forests. There are now hundreds of trees and thousands of saplings per hectare instead of a couple dozen or so big trees surrounded by a sea of native grasses. The large, mature ponderosas are well adapted to surface fires. Now, however, when fire gets into unnaturally overcrowded ponderosa forests, smaller trees act as a ladder for surface fires to enter the crown and become crown fires. Ponderosa have neither evolved nor adapted to contend with crown fires, so the entire forested ecosystem burns. It then takes decades for this ecosystem to recover naturally, because ponderosa seeds become limited. In some cases the ponderosa forests may not return at all, as parts of the West are becoming too dry for these forests to re-establish.

The third type of lightning-induced forest fire is a ground fire. When the forest floor and all its decomposing wood, fauna, bacteria and fungi are alight, it is a disaster. Such a fire can burn for months under a snowpack and flare up again in the springtime. This is a common problem in some higher-elevation forests and throughout British Columbia and Alaska. Ground fires sear the forest floor and deplete soil nutrients that have taken decades, and in some cases a century, to accumulate. One major concern about climate change is drought. Droughts promote the drying out of the forest floor, which in turn makes the forests increasingly prone to ground fires.

Fire is one of nature's important agents of change. Fire initiates a cycle of enrichment and fosters diversity of vegetation and micro-organisms. When fire is excluded from certain western ecosystems, it leads to a decline and the eventual disappearance of dominant shade-intolerant species like aspen, coastal Douglas-fir, ponderosa pine, sugar pine, western larch and western white pine. The loss of these species is not only tragic, it is changing the ecosystem interactions of our entire planet.

Humans have deliberately interrupted the fire cycle. As a result of immense fuel buildup in vast areas, large fires now burn almost twice as much area across the West as they did historically. As temperatures on the globe rise, forests are drying out and becoming prime candidates for large, raging firestorms that burn hotter due to more accumulated fuel. Forests are the life force of the planet, and human mismanagement is destroying their very existence and therefore the myriad ecosystem services they provide to all life.

Global Warming, A Climate Disrupter

As a conservation biologist, I am charged with
the responsibility of helping to maintain Earth's
breathtaking tapestry of life. In order to begin
to understand wild forests and their superb
complexity, it is essential to recognize that all
life has adapted to a range of habitable temperatures, and, importantly, that all life is intercon-
nected. As temperatures rise, the effects amplify
throughout the biosphere.

As a field scientist, I am trained to observe patterns in nature, patterns that help me understand
how biological systems function. Every climate
model I am aware of has predicted that Earth's
temperature will rise by at least 1°C (1.8°F), but
likely between 2 and 4°C (3.6 and 7.2°F) by 2100.
Potential consequences of the mean global temperature rising by these seemingly small amounts
are remarkably concerning. As you read, consider

that many of the changes described here have occurred over the past 40 years, during which time
the increase in Earth's mean global temperature
was just a little over half a degree (just under 1
degree Fahrenheit).

According to the World Meteorological
Organization, the ten warmest years on record
have all occurred since 1998. During this time,
extreme rainstorms and snowfalls have grown
substantially stronger. Human-caused greenhouse gas emissions have reportedly triggered
ecological changes in many ways: temperature
extremes, humidity levels, rainfall amounts,
ocean heat content and salinity, sea level pressure, increased wind speeds and wave heights on
the oceans, decrease in sea ice and snowpacks,
changes in the timing of runoff, and wildfire
damage. The combination of these changes, in
various times and places, is dramatically changing the patterns and behaviour of all life on our
planet.

The last time CO_2 levels were this high – 392
parts per million – was 15 million years ago, when
global temperatures were between 3 and 5.5°C
(5.4 and 9.9°F) higher, with sea levels between 23

and 36 metres (75.5 to 118 feet) higher. We know this from the work of scientists studying the ratio of the chemical element boron to calcium in the shells of ancient single-celled marine algae. Incidentally, according to researchers, during this ancient era there was no permanent ice cap in the Arctic and very little ice on Antarctica and Greenland. There is plenty evidence that CO_2 is linked with environmental changes in terrestrial ecosystems, ice distribution, sea levels and monsoon intensity.

Scientific work delving back a couple of million years into the Paleolithic found that the Arctic reliably amplifies by a factor of three whatever the global climate is experiencing. During the winter of 2010, tiny-shelled organisms called foraminifera revealed that, 1500 metres (4,921 feet) below the surface, the water flowing from the North Atlantic into the Arctic is 2°C warmer (3.6°F) than it has been in the last 2,000 years. Currently the Arctic ice is melting at a phenomenal rate. Thirty years ago there was 999,000 square kilometres (385,716 square miles) of thick, at least five-year-old ice in the Arctic. In September 2010, only 57,000 square kilometres

(22,000 square miles) remained – a decrease of 942,000 square kilometres (364,000 square miles), or 94 per cent.

As humans contribute to global warming by burning fossil fuels at unprecedented rates and destroying forests that act as carbon sinks, the resulting higher global temperatures are also creating the necessary conditions for the Earth itself to release the carbon it has been storing. In February 2011, the US National Oceanic & Atmospheric Administration (NOAA) and the National Snow & Ice Data Center (NSIDC) presented unequivocal evidence that the thawing permafrost will turn the Arctic from a carbon sink into a carbon source in the mid-2020s. That is just over a decade away. The frozen soils of the Arctic contain 1.5 trillion metric tons of frozen carbon, about twice as much as is currently in the atmosphere. Furthermore, much of the carbon will be released as methane, which is 23 times stronger at trapping heat than CO_2.

No current climate model has accurately incorporated this mega-addition of more than 1 billion metric tons of carbon per year into the atmosphere by the mid-2030s. NSIDC scientists estimate that "by 2200 an extra 190 gigatons plus or minus 64 gigatons of carbon" will enter the atmosphere. As climate scientist Kevin Schaefer, the lead author of the study, put it: "That's roughly equivalent to half of the total fossil fuel emissions since the dawn of the Industrial Age. Or to put it another way, it's equivalent to the emissions from all power plants in the United States for 80 years." With the thawing underway and the subsequent carbon that will be added to our environment – along with the deforestation and destruction of biodiversity on the Earth – we *need* to start answering the calls to lower our fossil-fuel emissions now.

In the meantime, the northern hemisphere – in particular North America – has experienced two brutally cold winters (2010 and 2011) compared to the previous four decades, with as many as ten significant snowstorms across the Midwest, the eastern seaboard and the southeastern US. Paradoxically, these storms are inexorably linked to soaring summertime temperatures in Greenland and northern Canada (as much as 6°C, or 10.8°F, above normal). These temperatures are rapidly melting summer Arctic sea ice, which results in smaller and smaller September ice covers to grow winter ice. Ice in the Arctic and elsewhere is vital for a number of reasons, but primarily it helps to reflect incoming solar radiation, which keeps our planet within a habitable range for humankind and sustains our intensive agriculture.

Around Christmas 2010, a colleague of mine called and informed me that it was raining on the southeastern coast of Baffin Island, in Iqaluit, the capital city of Nunavut – the vast eastern-Arctic territory of Canada's Far North. On January 4, 2011, the temperatures around south Baffin Island reached record highs as much as 22°C (40°F) above normal. Heat stored in the ice-free Arctic Ocean is being released into the atmosphere. Scientists have tagged narwhals – medium-sized toothed whales that live year-round in the Arctic – with sensors to record ocean depths and temperatures during feeding dives. With the narwhals diving as deep as 1773 metres (5,817 feet), the temperature sensors on them indicated an increase of 1 degree above normal. To raise a pot of water 1 degree on the stove is an easy feat, but to raise the temperature of the Arctic Ocean 1 degree is a daunting challenge with extreme consequences for all life.

In December 2010, an immense amount of ice – equal to the area of California, Oregon, Washington, Idaho and Nevada combined – was missing from the region. Still, in January 2011, a vast area of the eastern Arctic waterways was abnormally ice-free, and neither Frobisher Bay nor Davis Strait had frozen over entirely. Usually, at least half of Hudson Bay is frozen over by the end of November, but in early January 2011 only 17 per cent was frozen. Latent heat was fuelling the Arctic heat wave and high-pressure system that impeded normal freeze-up. Normally, a low-pressure cell resides in the Arctic during the winter months, but missing Arctic sea ice in the winters of 2009–2010 and 2010–2011 created a high-pressure system and frigid polar air has spilled as far south as southern Florida and Mexico. Along with a cornucopia of other implications, wild-weather patterns of a warming world are a major disruption to agriculture worldwide.

Earth is warming so quickly that early in the 2020s all subtropical and tropical glaciers in China, Indonesia, Mexico and the Andes will be gone. In April 2010, in Peru, a colossal chunk of ice measuring 497 by 199 metres (1,630 by 650 feet) broke off from Hualcan Glacier and plunged into a lake near the town of Carhuaz. The subsequent 23-metre (75-foot) wave killed three people and destroyed a water processing plant that served 60,000 local residents. Then, in August 2010, Greenland's Petermann Glacier lost a chunk of ice four times the size of the island of Manhattan. As the ice disintegrated into Davis Strait, it kept the iconic narwhals from surfacing in parts of their summer breeding grounds. The sad fact is that both Greenland's and Antarctica's ice sheets are actually shrinking significantly quicker – a year-over-year acceleration rate three times greater – than predicted in 2007 by the United Nations Intergovernmental Panel on Climate Change. From monthly satellite surveys between 1992 and 2009, researchers have determined that the two regions lost a combined average of 36.3 billion metric tons more ice each year than the previous year.

The ice sheets of Greenland and Antarctica will dominate future sea-level rise because they hold so much more ice mass than all the mountain glaciers combined. According to Richard Alley, an eminent geoscientist at Pennsylvania State University and host of a PBS television program on climate change called *EARTH: The Operators' Manual*, "What is going on in the Arctic now is the biggest and fastest thing that nature has ever done." Addressing a US House of Representatives committee on energy independence and global warming, Alley warned that "sometime in the next decade we may pass that tipping point which would put us warmer than temperatures that Greenland can survive." A rise in the range of 2 to 7 degrees would mean the obliteration of Greenland's ice sheet, which would result in sea levels rising 7 metres (23 feet). Coastal areas around the globe would end up under water.

Before it is too late, glaciologists are racing around the globe in an attempt to collect ice cores from the remaining glaciers, an important source of data about historical changes in climate. Part of the technique consists of measuring seasonal deposits of atmospheric dust on the ice, similar to measuring tree rings. Also, trapped oxygen within the ice clearly shows scientists differences in temperature over time, enabling them to understand how ancient weather changed. But the glaciers that are such an important source of this kind of historical climate data are disappearing. Indonesia's last remaining one, on Puncak Jaya mountain, has lost about 80 per cent of its ice since 1936, with nearly two-thirds of that loss having occurred just since the early 1970s. Scientists predict that the rest of this final Pacific region glacier will melt within a few years, taking with it evidence of climate history in the area. Another such icefield, on Iztaccihuatl, a dormant volcano that can be seen from Mexico City, will be gone by 2015.

The southern hemisphere shows further glaring indications that the burning of billions of tons of fossil fuels, and the greenhouse gases this adds to the atmosphere, is causing runaway human-forced climate change. Furthermore, the natural systems that have evolved to absorb CO_2, such as the forests and the oceans, are dying and releasing vast amounts of additional CO_2 themselves. Our world is warming and there is evidence all around us. Eight ice shelves have partially or fully collapsed along Antarctica's crooked 1450-kilometre (900-mile) finger of land that juts toward the tip of South America, where winter temperatures have increased by 6°C (10.8°F) and annual mean temperatures have risen by 2.7°C (4.9°F). Of the 244 glaciers along the western Antarctic Peninsula, 90 per cent have retreated since 1940. A small colony of emperor penguins on Emperor Island off the Antarctic Peninsula is gone and the species' future on the Antarctic continent is predicted to be bleak by mid-century. Some Adélie penguin colonies have also already gone extinct due to the latent heat content in the waters.

The seas absorb about 25 per cent of the world's emissions of CO_2, where it is converted to carbonic acid. The oceans' pH value, a scale from alkaline to acid, has fallen more than 30 per cent in a dramatic shift to acidity since the Industrial Revolution. The oceans are experiencing the fastest shift in chemistry in 65 million years, which will certainly impinge upon the beleaguered commercial fishery stocks and threaten the very existence of shellfish like mussels, shrimp and lobsters. Along with increased acidity of the waters, warming surface temperatures have lambasted Earth's coral reefs and their rich array of biodiversity. Warming surface water is bleaching many coral reefs and retarding the growth of others. Furthermore, in the warming Atlantic Ocean over the past 10 years, there have been three times as many Category 5 hurricanes as in the previous decade. In the past six years, five Category 4 or Category 5 cyclones have roared over Australia's Great Barrier Reef, causing substantial damage, while there had been only two of that ferocity in the previous 40 years.

As coral reefs are harmed or destroyed, we lose the potential to discover the powerful medicines that come from them. For instance, researchers have discovered that the toxic venom from the poisonous *Conus magus* sea snail that lives in the coral reefs in the Philippines is a naturally perfect compound for treating pain. Synthetically recreated, this compound is now the blockbuster pain medication Prialt (ziconotide), which is 100 times stronger than morphine and – unlike morphine – considered non-addictive. Beyond that, soft corals from northwestern Australia are the most efficacious anti-cancer compounds ever found, and Caribbean sea squirts are used to treat melanoma and breast cancers. Since 1969, sponges from the Florida Keys have played a valuable role in treating leukemia, and it was research into these that led scientists to develop the important AIDS drug AZT. Coral is the most effective treatment in regrowing human bones, with patients requiring no immunosuppressing drugs. Incidentally, ocean-derived pharmaceuticals are so important that Merck, Lilly, Pfizer, Hoffmann-La Roche and Bristol-Myers Squibb have all established marine biology divisions. The value of coral reefs in terms of ecosystem services has been estimated at $700-billion annually. Coral reefs are our children's legacy.

Global warming is also having a huge impact on Earth's fresh-water lakes. NASA scientists examined 104 large inland lakes around the globe and found, on average, they had warmed at least 1.1°C (nearly 2°F) since 1985. That is about two-and-a-half times the increase in global temperature in the same time period. In addition, Earth's rising, human-caused greenhouse gases appear to be forcing the water cycle to accelerate, boosting the amount of water flowing from the rivers into the oceans. Scientists from the US and India found that between 1994 and 2006, rivers discharged 18 per cent more water. As Earth warms up, more water is discharged from melting glaciers into the oceans and the warmer atmosphere can hold more evaporated water. From that point, once there is more water available, there will be more rain.

Climate-driven intense storms, flooding and heavy rainfall will put areas around the world, including the USA and Western Europe, in jeopardy. An increase in rainfall will cause aging sewer systems to overflow bacteria, viruses and protozoa into drinking water and onto beaches. Rising temperatures also pose a significant hazard to people living in areas exposed to waterborne diseases. Already the bacterial pathogen cholera has been on a record pace since its ongoing seventh pandemic outbreak, which began in Indonesia in 1961. Since that outbreak, cholera has spread into 50 countries and 7 million people. Researchers in turn have gained new understanding of the disease and the fact that as water and atmospheric temperatures continue to increase, *Vibrio cholerae* – the bacterium that causes cholera – will grow more rapidly in the warmer waters. This combination of factors means that outbreaks of infectious diseases like cholera will dramatically increase in frequency.

Every year since 1997, phytoplankton – the basis of the entire Arctic marine food chain and a crucial absorber of CO_2 – has experienced its peak bloom earlier in June than usual. Blue whales time their migratory 6400-kilometre (4,000-mile) trip to coincide with these blooms, gorging for up to 18 hours a day in order to return to warm tropical waters to breed. Other land animals are completely out of step with how quickly springtime is now arriving. For example, a 17-year study in the Rocky Mountains found that global warming is melting snowpacks three weeks earlier than the historical normal time, which is causing glacier lilies to emerge at least a couple of weeks before they are due. This is causing a climate-driven mismatch whereby bumblebees (*Bombus occidentalis* and *B. bifarius*) that pollinate the lilies are waking up two weeks too late. Without pollination, the fate of the lilies is uncertain. Furthermore, if you are one of the millions who suffer from allergy season, you will be rather disheartened to know that since 1995 the season is getting significantly longer. For instance, in Papillion, Nebraska, allergy season is longer by 11 days; in Minneapolis, Minnesota, by 16 days; in Winnipeg, Manitoba, ragweed season lasts 25 days longer; and in Saskatoon, Saskatchewan, it is a whopping 27 days longer.

The warming world is also responsible for the significant increases in food prices at street markets and supermarkets around the world. In one month, February 2011, the cost of vegetables increased by 50 per cent in the US. The last time US food prices rose that much was in 1974, during the first "oil price shock." All staple crops are at record prices on the world's markets because temperature, humidity, intense rainfall and prolonged drought are occurring around the globe. In India, for instance, in the state of Assam, tea production fell from 511,000 metric tons a year in 2007 to 417,000 metric tons in 2010. However, it is not just output that is of concern. Damp weather is unfavourable for tea, and in Jorhat, the main tea-growing area in Assam, they are experiencing record dampness, which is promoting insect attacks on the tea crops. Three million people in India alone work in the tea industry, most already living barely above the poverty line.

The same story is mirrored on the other side of the Earth in Colombia, except there the crop is coffee. Since coffee is one of my favourite vices, I've followed this story very closely. In 2006 Colombia produced more than 12 million 40-kilogram (88-pound) sacks of coffee and set an ambitious goal of 17 million for 2014. In 2010, the yield slid to 9 million sacks. Cultivation of high-quality coffee plants depends on a fine balance of heat, cold, moisture and dryness. At higher temperatures, coffee buds will often abort or the fruit will ripen too quickly. Also, coffee rust, a devastating fungus that could not survive the formerly cool mountain weather is now thriving in the warmer and wetter environments. Another increase of 1 degree brings with it the potential to destroy the multibillion-dollar coffee industry for Colombia and other coffee-growing countries. With everyone from Folgers to Starbucks already increasing prices to contend with higher wholesale costs brought on by lower yields and higher demand, our desire for a daily cup or two may not be able to stand up against the potential upcoming challenge of what some are calling "peak coffee." Along with coffee, prices of every commodity will continue to rise. We just cannot know by how much.

New weather patterns and the dramatic rise in CO_2 in our atmosphere are combining to unravel not only our staple crops but also our wild ecosystems around the globe. In 2009 the International Union of Forest Research Organizations came to a very bleak conclusion: "The carbon storing capacity of Earth's forests could be lost entirely if the planet heats up 2.5°C [4.5°F] above pre-industrial levels." So far, we have increased by about 1.1°C (nearly 2°F), which means we are already well on our way toward this fateful threshold. The result of crossing it would be a hardly recognizable world.

Around the globe, forests are suffering the effects of a warming Earth through increases in the number and area of wildfires over the past decade, including fires burning in the Far North. The treeless, permanently frozen land between the Arctic icecap and the treeline is called tundra. As Arctic temperatures rise, climate-driven tundra fires also increase. In September 2007, the Anaktuvuk River fire burned more than 1000 square kilometres (386 square miles) of tundra on Alaska's North Slope, doubling the area burned in that region since recordkeeping began in 1950. Sediment cores from the burned area showed without a doubt that this was the most destructive fire at this site in at least 5,000 years.

Raging wildfires in the summer of 2010 in central Russia, Siberia and western Canada created an enormous cloud that was detected from space by NASA's Aqua satellite. The satellite, equipped with an atmospheric infrared sounder, found high concentrations of poisonous carbon monoxide at an altitude of 5.5 kilometres (3.4 miles) above sea level. Over 15 million hectares (37 million acres) in Russia and Siberia alone burned, releasing 700,000 metric tons of carbon monoxide each day. These fires in Russia and Canada formed a ring of pollutants that circled the planet.

Devastating wildfires such as these are occurring all across the world. Even drought-laden Israel experienced its worst wildfire in modern times, with over 47 square kilometres (18 square miles) of the "lung" of the country, the Mt. Carmel Forest, being blackened in December 2010. Across North America the combination of diminishing snowpacks melting earlier and longer dry seasons are creating greater opportunities for large fires due to both the longer period in which ignition can occur and increased dryness of soils and vegetation. In the early spring of 2011, wildfires were alight in tinder-dry Colorado, New Mexico, Georgia and Texas – just to name a few.

Rising greenhouse gases are also wreaking unimaginable havoc in the tropical forests, more specifically in the Amazon. The heart of the Amazon has not evolved to contend with winds, never mind fierce winds, nor with drought. In 2005 a vicious combination of climate disruption occurred across a 1.9 million square kilometres (733,600 square miles) of land. In January an intense thunderstorm, spanning 100 kilometres long by 200 kilometres wide (62 by 124 miles), ripped through the whole Amazon Basin from southeast to northwest. On its path, the storm levelled between 441 million and 663 million trees – or the equivalent of 23 per cent of the estimated mean annual carbon accumulation capacity of the Amazon forest.

Later in 2005 a "one-in-one-hundred-year" drought occurred. Not only did the Amazon fail to absorb 1.5 billion metric tons of CO_2 that year, but over the next decade it will release approximately 5 billion metric tons of CO_2 from decomposing trees. If that isn't alarming enough, another mega-drought occurred across 3 million square kilometres (1.16 million square miles) in the Amazon in 2010, the second once-in-a-hundred-years event in five years. The enormous swath of dead jungle will release 8 billion metric tons of CO_2 over the next decade. And if the Amazon forests die, the Earth also loses its amazing cloud-making machines and will be forced to absorb incoming solar radiation rather than reflect it.

In 2009 the USA alone emitted 5.4 billion metric tons of CO_2 from fossil fuel use. These emissions contribute to an equally disconcerting worldwide pattern that is beginning to emerge. Scientists have documented that greenhouse gas emissions have significantly altered global climate – increasing the frequency, duration and/ or severity of drought and heat stress in 88 forests on every wooded continent on planet Earth. If ever there were a clarion wake-up call, this is it, without exception. All forest types are suffering from a deadly combination of at least three factors: insects and diseases associated with elevated temperatures; the drying out of plants; and carbon starvation, that is, water-stressed trees are unable to photosynthesize, or make food. Every decade since 1970 has seen more than a tenth of a degree of additional warming, which has caused less snowfall, declining snowpack water content and longer summer drought periods. Both old and young trees are affected.

Forests are changing all over the globe. Extreme droughts in North Africa are killing Atlas cedar from Morocco to Algeria. Heat and drought are battering the high-elevation tropical moist forests in Uganda, mountain acacia in Zimbabwe and centuries-old aloe plants in Namibia. Tropical forests of Malaysia and Borneo have also suffered significant death. Drought has also lambasted the tropical dry forests of northwest and southwest India, fir in

South Korea, the junipers of Saudi Arabia, and pine and fir in central Turkey. Extensive areas of forest in southwestern and east-central China have now been recognized as being at a high threat of mortality in the ensuing years. Russia too has identified 76 million hectares (187.8 million acres) of high-threat southern forests, where drought is severely stressing trees. Australia has seen widespread death in acacia woodlands and eucalypt and *Corymbia* forests. New Zealand has documented drought-induced death in high-elevation beech forests. Oak, fir, spruce, beech and pines across western Europe are dying too.

Across western North America the western spruce budworm (*Choristoneura occidentalis*) has been on a binge in the conifer forests. Rising temperatures have paved the way for this formidable defoliator to enter, for the first time, the interior of Alaska and the Yukon, both of which formerly were too cold for the budworms to exist. Warm temperatures throughout Alaska have also allowed an invasive, non-native sawfly to kill green alder – wonderful nitrogen-fixing, fertilizer trees – in the south-central parts of the state. Alders are keystone species that provide food and support the health of entire ecosystems. For instance, they enrich soils, feed moose and prevent riparian banks or edges from eroding. Along with the alders, another keystone species has also begun to die with vengeance across the western USA. In 2005, scientists in Colorado discovered 13,600 hectares (33,600 acres) of quaking aspen dying en masse. The next year, 68,000 hectares (168,000 acres) of the aspen died. By 2008 the death toll had leaped to 251,000 hectares (620,000 acres). Warmer temperatures and dry weather have proven again to be lethal for these remarkable trees.

Nature is incredible. Trees are dying, but some appear to be putting up an evolutionary fight. Teams of ecologists from the Netherlands and the US have discovered that plants are beginning to respond to rising atmospheric CO_2 by significantly reducing the number and size of their stomata – the CO_2 exchange mechanisms on leaves. This is a remarkable attempt by the trees to adapt in order to survive the coming drier times. Researchers have also found that this very recent adaptation also results in a "huge" reduction in the release of water vapour into the atmosphere. While I would rather not speculate on potential outcomes of these changes, they distinctly show that species are being forced to adapt to our warming world.

Humans, on the other hand, are, as a species, vastly unprepared for the coming decades. In its annual statistical review for 2010, the Center for Research on the Epidemiology of Disasters, at L'université catholique de Louvain in Brussels, found that 385 natural disasters that year killed more than 297,000 people (including 56,000 from the intense and prolonged Russian heat wave) and caused in excess of US$123-billion in damage. As the atmosphere heats up, the forests continue to decline and the wild weather patterns widen their scope of destruction, humans will also have to learn how to adapt and contend with the new challenges brought on by climate change.

the good
ingredient
ne, for ir

Lodgepole Pines

Lodgepole pines are incredible trees. Three main races survived the Pleistocene glaciation – the current ice age – by migrating south down the Pacific coast, down the chains of the Rocky Mountains and down the California's Sierra Nevada range. Today, lodgepole pines are found from sea level to an impressive subalpine elevation above 3300 metres (10,827 feet), across 27 million hectares (66.7 million acres) of forested lands from the Yukon to California, east to New Mexico, back north in Nebraska, South Dakota, the Cypress Hills in southwestern Saskatchewan, across Alberta and up into the southwestern corner of the Northwest Territories.

Lodgepole pines are wonderfully adapted to fire. In fact, the Pacific and Rocky Mountain races possess cones that remain shut tight by resin that melts in the heat of fires. This type of release ensures that an abundant seed source is able to quickly colonize any mineral-exposed, post-fire soil and regenerate a forest of trees that are all the same age – or as foresters call it, even-aged. On average the trees can live for about 150 years, though some have been found as old as 200 years. Typically, however, trees live for about 75 years – sometimes substantially less – before lightning-induced fires reset nature's biological clock, kick-starting a new lodgepole pine forest.

These wild forests, like all their counterparts around the globe, are important for a number of reasons. Across western North America, for instance, they act as massive snow fences that hold winter snowfall and release the spring meltwater slowly. Their efforts help to feed streams, rivers and watersheds. In turn, they are essential for drinking water and agricultural irrigation for over 50 million residents across the West.

The species' common name, "lodgepole," derives from indigenous North Americans' use of these tall, straight trees to build dwellings, or lodges. The natives also used the inner bark, especially in the springtime, with its rich supply of vitamin C and other life-sustaining carbohydrates, as an essential food source. Moreover, the gooey lodgepole pitch, or sap, was a prized ingredient for many medicines. In the springtime, for instance, lodgepole pine buds were chewed on for effective relief of sore throat.

Much to the chagrin of most plantation foresters, but the delight of all naturalists, porcupines, the second-largest indigenous rodents in North America, also have a penchant for the rich inner bark of the lodgepole pine. Porcupines have unmistakable armour made of stout, sharply barbed quills, and oval, knobby soles – unique among North American mammals. Their powerful tail is their main defence, as their quills detach on contact. During the autumn and winter, porcupines depend upon the inner bark of these trees. It takes about 100 mature lodgepole pines to sustain one 18-kilogram (38-pound) porcupine over a winter.

Insects and animals such as porcupines use lodgepole pines for sustenance, but they do not destroy forests in the process. Sadly, this is not always the case. Currently, if you fly or drive anywhere across western North America, you will encounter massive swaths of red-brown, dead-topped conifers. Something, even to the untrained eye, appears to be very wrong across the entire West. That something is native, tree-killing bark beetles. In order to fully understand what is happening across the West, keep in mind that trees, insects and the climate are all inextricably linked; each plays a pivotal role in the feedback loops on nature's gameboard. Feedback loops help to maintain ecosystem stability. When disrupted, however, even seemingly small effects can be amplified to very large scales. Any change in the behaviour of one or two of the players inevitably changes all of the triumvirate's interactions along with it. With little opportunity to adapt to the new conditions, instability can cause remarkable devastation to the whole ecosystem.

Traditionally, wildfires and bark beetles are among nature's greatest agents of change in the forest. Currently, a collision of human-created events has encouraged conditions for a perfect storm, one that allows the beetles to embark on a feeding frenzy never witnessed before. A hundred years of fire suppression in human-managed forests across the West, combined with rising temperatures and region-wide droughts, has dramatically altered the ecosystem's stability feedback loops and unleashed hundreds of billions of voracious bark beetles. Bark beetles have become, by far, the leading force of change in forests across western North America.

Of the 1,400 native bark beetle species across the region, only a handful – less than 1 per cent – actively kill trees. In biological classification terms, they all belong to one family called *Curculionidae*, one subfamily called *Scolytinae*, and three genera: *Dendroctonus*, *Ips* and *Scolytus* (see table below). Tree-killing bark beetles are thought to have evolved with conifers a couple hundred million years ago during the Triassic Period. There is evidence of the genus *Dendroctonus* from larch amber in the Canadian high Arctic from the mid-Eocene, or about 45 million years ago. Bark beetles have also been found in fossil remains in the Rocky Mountains, in Colorado, near Snowmass Village, dating back 150,000 years. More recently, there is evidence of their existence during the last glaciation, the Pleistocene, from pine bark samples found in packrat middens in New Mexico and tar seeps in the La Brea Tar Pits in Los Angeles. These show that bark beetles were on the scene about 30,000 years ago. Remains of *Dendroctonus* from 8,000 years ago were also discovered in lake sediment cores from the Bitterroot Mountains in Montana.

The mountain pine beetle (*D. ponderosae*) is the most destructive native pine beetle in western North America. Its range, before 2000, extended from northern to central British Columbia in Canada and from the Pacific coast to the Black Hills of South Dakota in the United States. According to a 2010 study published by the Entomological Society of Canada, the range has yet to encompass the northern reaches of its main host – lodgepole pine – but this is changing quickly. Mountain pine beetles principally attack and kill lodgepole pine, but also western white pine, ponderosa pine, whitebark pine and limber pine. In arboretums – botanical gardens devoted to trees – they can also attack eastern white pine, pitch pine, red pine and jack pine. So far, only the Jeffery pine appears impervious to them. Mountain pine beetles can also easily attack true firs and spruce, although successful reproduction has yet to be confirmed in either genus.

Mountain pine beetles, like all other destructive beetles, govern every stage of their life according to temperature thresholds. Temperature determines development, survival and reproductive success. There are four life stages: egg; larva with four instars, or stages; pupa; and adult. Mountain pine beetles are especially designed in the third or fourth instar to tolerate extreme temperatures of −40°C (−40°F) in the winter. In order to supercool their "blood" (the fluid in the circulatory system of some arthropods is correctly called hemolymph) they manufacture a polyhydric alcohol called glycol, which is also one of the main components in antifreeze fluid for automobile radiators. Glycol blocks ice formation by lowering the fluid's freezing point and preventing the growth of ice crystals.

Members of the *Pinaceae* family, including pines, spruces and others, possess a primary defence mechanism made up of a resin system. When an insect bores into the tree, the tree releases a sufficient flow of resin, or pitch, to preclude the insect from going any further. The phenomenon is called pitching-out. In addition, these trees have a secondary defence system, which specifically responds to wounds.

Table: Biological classification of bark beetles that actively kill trees in western North America

Taxonomic hierarchy (general)	Taxonomic classification of bark beetles	Special characteristics and notes
Domain	*Eukaryote*	A cell or organism that possesses a clearly defined nucleus.
Kingdom	*Animalia* (animal)	Also called metazoa, these are multicellular and heterotrophic (relying directly or indirectly on other organisms for nourishment). Most ingest food and digest it in an internal body cavity.
Phylum	*Arthropoda* (arthropods)	An invertebrate animal with an external skeleton, a segmented body and jointed appendages.
Class	*Insecta* (insects)	An animal with a chitinous exoskeleton, a three-part body (head, thorax and abdomen), three pairs of jointed legs, compound eyes, and two antennae. This class includes more than a million species, representing more than half of all known living organisms.
Order	*Coleoptera* (beetles)	Contains more species than any other order – almost a quarter of all known forms of life.

Family	Curculionidae ("true" weevils and snout beetles)	Over 40,000 plant-eating species worldwide, one of the largest families.
Subfamily	Scolytinae	From the Greek skolytein, which means to cut short, perhaps alluding to the cut-off abdomen of some species.
Genus (plural: Genera)	Dendroctonus	Dendroctonus is the most aggressive and economically important genus in this subfamily.
	Ips	Ips is generally considered less aggressive than Dendroctonus but frequently builds up populations in dead wood and can then attack and kill living trees.
	Scolytus	One member of genus Scolytus, the European elm beetle, is a carrier of the dreaded Dutch elm disease that has devastated elm trees in many parts of North America and Europe.
Species (examples that are involved in forest destruction)	Dendroctonus ponderosae	Scientific name (genus and species) for the mountain pine beetle.
	Dendroctonus brevicomis	Scientific name for western bark beetles.
	Ips typographus	Scientific name for European spruce bark beetle.
	Ips spp.	Scientific name for engraver beetles.

Essentially the tree isolates the damaged area to cut off any food source to the invading organism. The resin contains monoterpenes, diterpene acids and stilbene phenolics, which are known to have antibiotic and other repellent properties to fight beetles and fungi, bacteria and mites.

When members of the *Pinaceae* become stressed by drought, their primary defence system – copious amounts of gooey pitch – is drastically compromised and they attract insects, particularly but not exclusively bark beetles. Bark beetles are dark in colour and about half the size of a skinny popcorn kernel. They are able to detect higher plant-surface temperatures, leaf yellowing, increased infrared reflectance, biochemical changes and likely stress-induced cavitation when cells desiccate. One method they use to find the drought-stressed trees is through their attraction to the ultrasonic acoustic emissions that result from breakage of water columns. For inter-beetle communications, they initially use scent, or pheromones, but recent research suggests that once beetles have colonized the tree, bioacoustics are also used as a secondary form of communication.

Bark beetles have evolved a clever predatory

strategy of turning the plant's terpenoid defence system against itself. First, the beetles overcome mature trees with a synchronous mass attack. Sheer numbers completely overwhelm the stressed tree, or even a healthy one. The tree emits chemicals – known as kairomones – that actually attract more invaders. Initially in August, however, it is the pioneer female mountain pine beetle that penetrates the bark and oxidizes the tree's monoterpene alpha-pinene – a type of terpene – to produce trans-verbenol, a pheromone that attracts males. Once the males arrive, they release a pheromone called exo-brevicomin at a low concentration to attract a female. When this has occurred, the tree's kairomones of alpha-pinene and myrcene blend together, and this signals that the tree has relinquished the fight. Basically, the tree itself sends out the signal that sparks a mass attack of thousands of mountain pine beetles, which usually lasts for about 48 hours. Once the tree is completely colonized – 60 attacks per square metre of bark – mountain pine beetles halt the attack by using yet another chemical cue as an anti-aggregate. It is hypothesized that high concentrations of trans-verbenol emitted from the attacked tree sends a signal to surrounding trees of impending death and acceptance of the forthcoming onslaught.

In order to defeat the tree's autoimmune system, the beetle has evolved a mutualistic relationship with ophiostomatoid blue-stain fungi and bacteria. There are two species of blue-stain fungus, *Grosmannia clavigera* and *Ophiostoma montium,* which are found throughout the entire range of the mountain pine beetle. A third species, *Leptographium longiclavatum*, appears to play a pivotal role on the east side of the northern Rockies, having adapted to the colder boreal forests of Canada. After the beetle bores through the outer bark into the tree's food supply – the rich sugar in the phloem tissue – spores of blue-stain fungus, bacteria and micro-organisms are released from their mycangia, which are specialized pockets in the beetle's mandibles (the pair of appendages near the insect's mouth). Quickly germinating, the beetle's partners effectively stop the tree's production of resin. They then proliferate, impairing the water-conducting tissue, or xylem cells, in the sapwood. In part, the trees perish from an inability to draw water up to the crown in order to manufacture food in their leaves by means of the elegant process of photosynthesis.

Once mated, the female beetle deposits about 65 individual 1-millimetre (0.04-inch) eggs into invaginations cut in the side gallery on the inside of the tree's bark. Two-thirds of the eggs are females. The larvae go through four growth stages, or instars, feeding on the tree's nutritious sugars in the phloem and on the fungus and bacteria that now line some of the phloem tissue. By late September the larvae have usually entered their fourth stage and they empty their gut to overwinter successfully and avoid freezing. The next spring, once temperatures have risen above 16°C (61°F), the beetles complete their maturation by feeding on inside bark and spores of blue-stain fungi and other micro-organisms. They strengthen their flight muscles, increase in size to about 6 millimetres (0.24 inches) in length, and fill their mycangia with fungal and yeast spores to infect and colonize new trees. Other micro-organisms are transported to new trees by means of sticky spores that hold fast to the beetle's exoskeleton.

During the conquest of the tree, the beetles are known to produce a host of noises, including stridulations, chirps and clicks, which go on for days and weeks after most of their other behaviours have apparently ceased. Bark beetles create friction-based sounds by grating a surface, probably using a stridulatory organ (*pars stridens*) on the back of the head (*Ips*) or under their wings (*Dendroctonus*). Bioacoustics in bark beetles is just beginning to be studied. It is likely that in addition to chemicals, sound signals also regulate aggression, attacks on host trees, courtship, mating behaviour and population density.

During dry weather, usually in August, as long as the ambient air temperature is above 16°C, but optimally 20°C (68°F), the beetles take flight. Temperature alone orchestrates the synchrony of the mass flights in these biblical-style attacks. The speed of flight, on average, is 2 metres (6.56 feet) per second. Some journeys are short range within a single stand of trees, whilst others are above the forest canopy at densities over 18,600 beetles per hectare (2.47 acres). Bark beetles can travel up to 110 kilometres (68.35 miles) a day, and synchronous attacks have been reported to occur at distances of up to 900 kilometres (559 miles).

Since the late 1990s in BC the mountain pine

beetles have killed an estimated 700 million cubic metres (nearly 297 billion board feet) of mostly lodgepole pine, a truly astonishing volume that would provide enough lumber to build more than 5 million new houses. The infestation is so enormous that the beetles have overcome tens of millions of healthy, non-stressed trees. At least 15.5 million hectares (38.3 million acres) of mature lodgepole timber, half of BC's commercial forests, are dead. Along with the multitude of other ecological, health and economic consequences of losing so much forest, we must begin to grapple with the eventual carbon output of the dying trees. What were huge, healthy forests that absorbed CO_2 and emitted oxygen will, over the next ten years, decompose and release 250 million metric tons of CO_2 into our atmosphere.

The vast outbreaks of bark beetles have occurred in BC, particularly over the past 15 years, due to the absence of lethal early fall and late spring hard frosts that would reach temperatures from $-25°$ to $-40°C$ ($-13°$ to $-40°F$). In addition, in 2002 and again in 2006, billions of mountain pine beetles were sucked into the upper atmosphere and spat out hundreds of kilometres to the east, on the east side of the Rockies in the region of Grande Prairie, Alberta. Millions of the beetles that experienced this forced migration not only survived but successfully reproduced.

Coincidentally, in 2006 my faithful Chesapeake Bay retriever, Naio, and I were exploring the woods near Grande Prairie when the sky rained billions of beetles on us. It felt like we were being pelted by grains of rice. The shower continued for at least ten minutes. When I figured out that it wasn't rice, but beetles, I was absolutely shocked. Fortunately, there was a big, old spruce hollow nearby, so Naio and I took shelter from the beetle bombardment.

Since 2006 the mountain pine beetles have noshed their way through Alberta's lodge-pole–jack pine hybrids in the northwest of the province and into the pure jack pine stands – an amazing adaptation by beetles advancing their way in a warming climate into a pine species that has never evolved to contend with such aggressive predation. Beetles are also success-fully breeding within a year, increasing their range of synchronous attacks – advancing east of Fort McMurray to within a couple hundred kilometres of Saskatchewan – and now the brood adults are significantly smaller than the parent adults. Within one generation, the ravenous bark beetle populations have adapted to and are reproducing in trees with thinner phloem and smaller diameters, with a minimum of 1.5-millimetre (0.06-inch) bark thickness.

Bark beetles are successfully adapting in so many ways that scientists predict that, as Earth continues to warm, jack pines in the boreal forests that extend across Canada to Labrador on the Atlantic coast will also become vulnerable to beetle infestations. The 1 to 2°C (1.8 to 3.6°F) increase in mean annual temperature in central BC since 1970 has paved the way for beetle populations to expand into more northerly latitudes and into the higher-elevation pine forests. Modellers have predicted that with a warming of 2.5°C (4.5°F), beetles will advance northward by seven degrees of latitude into new, climatically suitable habitat. The northern boreal forests, land that is currently *terra incognita*, will undoubtedly lose such a distinction for the mountain pine beetles.

In the meantime, the lodgepole pines in northeastern BC, formerly a range uninhabitable for the beetles, are now accessible. The northern pines are more chemically attractive to the beetles compared to the southerly lodgepoles. Northeastern lodgepole pines lack the terpenoid defence mechanism to fight the beetles, because they did not expend the energy resources to build such a defence system against a non-existent (at the time) beetle predator. As a result of global warming enabling the beetles to advance into these naïve forests, beetle populations are breeding with much higher reproductive success than in the southerly lodgepole pine forests.

Mountain pine beetles have also been on an epic all-you-can-eat smorgasbord in the US, across eastern Washington, Idaho, Montana, Wyoming, Colorado and South Dakota. They have devastated over 2 million hectares (4.9 million acres) of lodgepole in Colorado and Wyoming alone. Even in the most easterly edge of the lodgepole range, in the Black Hills National Forest of South Dakota, these beetles have reached epidemic levels. In many stands across the US, mortality rates are in excess of 90 per cent, as many smaller-stemmed trees have also been killed from intense infestations.

Tree-ring records dating back to 1750 reveal intermittent outbreaks of bark beetles, but over the past 15 years the frequency, severity and extent of bark beetle outbreaks has significantly increased across all of western North America. There is every reason to believe that jack pines, red pines and eastern white pines along the lake states in the northeastern United States will, with rising temperatures, also face mountain pine beetle incursions.

Over the past hundred years, lightning-induced fires have been deliberately suppressed across western North America. The frequency of natural fires in lodgepole forests is about one major, stand-replacing fire every 40 to 200 years. In an attempt to manage wild forests for timber production, foresters have inadvertently created an abundance of food for bark beetles. At the beginning of the current outbreak in BC, there were three to three-and-a-half times as much mature lodgepole on the landscape as during the previous hundred years. Moreover, beetle-killed trees retain their needles for up to five years, which contain highly flammable terpenes. The dead pines essentially become kindling for mega-firestorms, which have incinerated communities from San Diego to Kelowna and many in between over the past 15 years.

Since the Industrial Revolution, CO_2 levels have risen about 40 per cent. The natural feedback loops between trees, insects and the climate have changed irrevocably. Higher temperatures, particularly in the fall and winter, have significantly reduced mortality rates of bark beetles and allowed the temperature-governed beetles to synchronously feed at will on mature lodgepole pine forests, as well as jump into the jack pines in the Far North. Furthermore, warmer springs are melting snowpacks earlier and increasing the frequency and duration of wildfires across western North America. Concomitantly, the greatest increases in wildfires in the US since 1970 have occurred in mid-elevation northern Rocky Mountain lodgepole forests where bark beetle eruptions have been significantly increasing. Any subsequent temperature rise of between 1 and 2°C (1.8 and 3.6°F) will be endgame for many lodgepole pine forests; the species would likely survive in only 17 per cent of its current range of 27 million hectares (66.7 million acres).

The ecosystem services of lodgepole pines are crucial for absorbing rising CO_2 levels, regulating the water cycle and maintaining healthy soil ecosystems throughout western North America. Lodgepole pines provide habitat for so many critters, including apex predators like black bears and mountain lions, along with millions of song-birds. These forests offer species diversity, which enriches the genetic tapestry of life. Interrupting the fire cycle to create merchantable timber has inadvertently provided a food source for one of nature's formidable agents of change – the mountain pine beetles. These predatory beetles have significantly benefited from both rising greenhouse gas emissions and the associated elevated temperatures, which have enabled them to expand their range and reach historic population levels throughout western North America. The beetles' benefit is our loss, as we must now contend with the devastation of vital forest eco-systems in the aftermath of the ravenous beetle attacks.

Spruce Forests

Six incredible species of spruce, and one naturally occurring hybrid, spread across western North America, occupying high-elevation, coastal and high-latitude forests. These trees are champions, having adapted to both the salty bite of ocean spray and incredibly frigid temperatures with short growing seasons. Spruce bark beetles (*Dendroctonus rufipennis*) exist across most of the spruce range, usually at low levels. Cold temperatures at high elevations and high latitudes typically prevent beetle populations from reaching infestation or outbreak levels. Over the past two decades, though, something appears to have gone very wrong. Not only have there been massive beetle outbreaks far beyond any historical records, but the infestations are also occurring across larger, regional networks of forest – on a regional scale. Moreover, vast tracts of moisture-starved mature spruce have stopped growing. A warming climate is disassembling the far northern boreal forests at a pace that is truly alarming.

Cores from glacial ice and marine sediments going back millions of years show that the Arctic reliably amplifies the global climate threefold, which is exactly what is happening in Alaska, the Yukon and northern British Columbia. Temperatures in the near-Arctic forests of those regions have risen by at least 1.4°C (2.5°F) since 1970, compared with a global average temperature increase of around 0.5°C (0.9°F) in the same time. Warmer temperatures, particularly in winter, have enabled a host of insects that defoliate trees the rare opportunity to enter ecosystems once dominated by fatal, brutally cold temperatures. Warmer winters in Alaska, the southwest Yukon and northern BC are linked with stronger circulation around the dominant Aleutian Low, a semipermanent low-pressure region located near the Aleutian Islands in the northern Pacific Ocean. Circulation within the Aleutian Low sucks up air from the central Pacific around Hawaii and delivers it due north into Alaska. When the

Aleutian Low intensifies, as evidenced over the past couple of decades, immense amounts of heat are discharged to the Far North.

As winter temperatures have changed, so too have the summers become warmer in the interior of Alaska, the Yukon and northern BC. These warmer summers are associated with dominant high-pressure cells that coincide with the long days around the summer solstice, which can have over 18 hours of sunlight. The strength of these high-pressure cells has been intensifying, and as a result they have successfully reduced summer-time precipitation deep into the western boreal region. The subsequent moisture stress has been devastating to mature white spruce in these areas, not only stopping their growth but also shutting down the crucial autoimmune defence mechanism – gooey resin – that helps the trees fight the deadly spruce bark beetles.

Warming summertime temperatures have actually brought such intense droughts to Alaska's boreal forests over the past couple of decades that many lakes are drying up. The permafrost is also thawing, creating tens of thousands of ground depressions called thermokarsts. Even in otherwise cool environments like the treeline, it has warmed up so quickly – thus reducing moisture – that the trees have responded by reducing growth. Summertime precipitation has decreased across Alaska and southwest Yukon to the extent that trees are having a difficult time making food, or photosynthesizing. Whereas growth in northern regions has always been limited by low temperatures, low moisture has actually now trumped low temperature in adversely affecting growth.

From 2000 to 2009, drought in Alaska's boreal forest resulted in a fourfold increase in late-season burning compared to the previous five decades. Many of these are ground fires, which burn deep into the stored carbon of the forest floor and release greenhouse gases into the ever-rising atmospheric pool. Some ground fires burn all winter long and ignite larger fires the following spring. Twelve million hectares (29.6 million acres) have burned over the past decade, and conservative estimates predict an increase by 200 to 300 per cent over the next 50 years. Alaska's boreal ecosystem has switched from a carbon sink – actively pulling CO_2 from the atmosphere – to a carbon source whereby it is releasing more than it is absorbing.

The drying wetlands across Alaska and northern Canada are playing another significant role in ecological feedback loops by releasing other toxins when they burn as well. Large-scale coal use began during the Industrial Revolution, and coal is still an important source of energy around the world due to its abundance and relatively low economic cost. Burning coal is extremely harmful to the environment, though. Already, it has released tens of thousands of metric tons of mercury vapour into the biosphere. Mercury vapour is an awful neurotoxin that causes fatigue, tremors, vision disorders and brain, kidney and circulation damage in humans – especially children.

Over the years, wetlands across the Far North have been quietly accumulating tons of mercury vapour. Wildfires are now releasing a century's worth of banked toxicity in a single blaze. When the toxins rain back to the ground or low-oxygen environments in the ocean, they are consumed by bacteria and converted to methylmercury. In this form the toxins are "sticky" and hang around – bioaccumulate – in animals' bodies. Sharks, swordfish, mackerel, tilefish and tuna are already exhibiting very high levels of methylmercury, which is also hazardous – and potentially fatal – for humans because the insidious toxin is biomagnified to an even higher concentration once consumed by humans.

As perhaps you have anticipated, the combination of rising temperatures, particularly in the Far North, and prolonged droughts are not only stressing mature spruce, but have given spruce bark beetles the rare opportunity to speed up their lifecycles. They are feeding and breeding like never before recorded. Over the past decade and a half, extensive forests of greater than 3 million hectares (7.4 million acres) have been killed by spruce bark beetles, including areas in the Rocky Mountains, Alaska, the southwest Yukon, BC and the contiguous USA.

Tree rings tell scientists about past beetle outbreaks. Those trees that were spared from infestations or were too young and unsuitable (diameters less than 15 centimetres, or 6 inches) often noticeably increased their rate of growth after a beetle outbreak, because of the lack of competition and plentiful sunlight and water. This rapid pulse of growth is easily detectable in tree-ring width among a number of trees in a stand during the same period of time. In most cases, infestations kill about 70 per cent of mature trees. By matching overlapping patterns of tree-ring growth – a technique called cross-dating – bark beetle infestations have been well documented across western North America over the past 270 years.

Spruce bark beetles evolved to kill mature species of spruce, such as Engelmann, white, Engelmann–white hybrid, Sitka and Lutz. Black and Colorado blue spruce are far less frequently attacked. Engelmann spruce are high-elevation specialists, often found with subalpine fir and lodgepole pine. White spruce too can easily reside in the high mountains, but they are typically northern latitude residents. In BC, Engelmanns and whites produce a splendid example of a fit hybrid. Engelmann, white and their hybrid are well adapted to winter temperatures colder than −35°C (−31°F). Sitka spruce are marvellous examples of resilient west coast rain forest trees that range from northern California to Alaska;

they often reach 95 metres (312 feet) and live for a millennium. Lutz spruce are natural hybrids between white and Sitka spruce and are native to coastal northern BC and coastal southern Alaska.

Spruce bark beetles' lifecycle – from individual eggs, to larva with four instars, to pupa, to adult – extends from one to three years. Usually it takes them two years to develop. In Alaska, southwest Yukon and northern BC, when temperatures reach 14.5°C (58°F) in May, mature adults synchronously attack spruce. By mid-October all the eggs have hatched and entered either the second or the fourth instar. The larvae overwinter with a built-in diapause response, or dormancy mechanism, which, incidentally, the mountain pine beetles lack. Normally, the spruce beetle larvae awaken the following spring and undergo pupal stage, metamorphose into adults and then overwinter again, awaiting springtime dispersal.

Spruce beetle larvae can tolerate temperatures as low as −34°C (−29°F), whereas adults die when temperatures reach −26°C (−15°F). Hence, in most cases, the spruce beetle populations remain endemic due to the high mortality rate during the second overwinter. Occasionally populations reach outbreaks, and when greater than 40 attacks per square metre (10.76 square feet) of bark occur, even healthy spruce are unable to defend against the beetle and its accompanying partners the blue-stain fungus, bacteria nematodes and other mites. Spruce beetles lack the mycangia, or pocket near the jaw, that the mountain pine beetles possess, so they carry their partners either under their elytra (wing cases) or on their exoskeleton.

Spruce beetles have adapted to breed successfully in freshly wind-thrown spruce trees, whereas mountain pine beetles require live, standing, mature trees to reproduce successfully. Natural enemies of both beetles include many species of woodpeckers, clerid beetles, diseases and other, vertebrate predators. Interestingly, both male and female spruce bark beetles produce a wide array of sounds believed by ecologists to be an important communication tool. Universities such as Guelph, Simon Fraser and others are currently completing important research in this nascent communication field of bioacoustics.

Between 1989 and 2004, spruce bark beetles killed over 1.6 million hectares (3.95 million acres) of spruce along the Kenai Peninsula and Cook Inlet region of south-central Alaska. Tree-ring analysis indicates that a sustained period of five to six years of warmer summers not only weakened the trees significantly but also enabled beetle populations to soar in widespread outbreaks. The unprecedented run of warm summer weather in Alaska began in 1987. Tree rings from south-central Alaska also confirm that endemic spruce beetle attacks have occurred at least five times over the past 250 years, but only one local outbreak in 1870 neared the enormity of the 1990s tree mortality.

Warm temperatures, above 14.5°C (58°F) from late May to early June, on the Kenai Peninsula resulted in early adult emergence and increased breeding activity. The beetle larvae also had a longer season to mature. When the sugar-rich phloem tissue reached a temperature of 16.5°C (61.7°F) during either the first or second larval instar stages (mid-June to mid-July), spruce beetle larvae matured to adults in one year instead of two. Moreover, warm weather synchronized the larval development and adult emergence, at least doubling the number of spruce bark beetles simultaneously attacking and successfully killing mature spruce. Sustained warm temperatures were also pivotal in enabling local spruce beetle outbreaks to expand their range and regionalize.

Kluane National Park is located in the far southwest corner of the Yukon. It is home to the spectacular St. Elias range, which boasts the largest peak in Canada – Mt. Logan (5959 metres, or 19,550 feet) – and effectively blocks moisture from the Pacific Ocean. A meagre 265 millimetres (10.4 inches) of precipitation falls each year in the southwest Yukon, so even a slight shortfall of moisture will stress trees. Kluane is a region where frigid temperatures dominate and only white spruce, aspen and balsam poplar can survive. Before the early 1990s most of the damage to forests in the Yukon was from recurring periods of intense cold.

Tree ring records from the southwest Yukon show only two previous spruce bark beetle infestations and both were restricted to the immediate area. Logging debris left from the Haines road construction in 1942 exacerbated a small infestation in a stand of white spruce that reportedly spread over several square miles.

A second localized attack in 1977, of over 100 hectares (247 acres) of white spruce, was traced to the felled white spruce that were left at the construction site while building the Aishihik dam and diversion canal. Humans undoubtedly played a role in both of the previous infestations in the southwest Yukon.

The warm temperatures Alaska experienced in 1987 also encompassed the Yukon. In addition to high temperatures, most of the white spruce in southwestern Yukon received less than normal precipitation. With increasing transpiration rates – the evaporation of water molecules from plant surfaces – and thin soils unable to replace the moisture, hundreds of thousands of mature white spruce were thrust into drought stress. Drought severely curtails white spruce resin defence mechanisms, and the water-stressed trees emit chemicals that all bark beetles can easily detect. In the late 1980s, approximately 25 per cent of the spruce bark beetles in southwestern Yukon began speeding up their breeding cycle from two years to one year. To make matters worse, the typically deadly temperatures in late November (below $-24°C$, or $-11°F$) did not occur, so overwinter survival rates of bark beetles were high.

By 1999 immense outbreaks began to occur in the Shakwak Trench and the Alsek River corridor in Kluane National Park as well as south toward the Tatshenshini River valley in BC. All spruce beetles in the region at this time were successfully breeding within one year. Bark beetle movement was aided by strong west winds spilling into valley corridors. Two successive growing seasons with unusually cool, wet weather in 2000 and 2001 caused spruce beetles to switch back to a two-year breeding cycle. Rehydrated trees provided an abundant food source, and by 2002 white spruce were killed over an area triple that of the previous year. By 2005 spruce bark beetles had accomplished something never recorded in modern or historical times: they had killed more than 350,000 hectares (864,870 acres) of mature white spruce in southwestern Yukon.

Rising temperatures in conjunction with regionalized spruce beetle outbreaks across the southwest Yukon appear to be causing irreversible ecosystem regime shifts. Spruce forests historically are able to regenerate themselves, but the spruce bark beetles have killed most of the seed-bearing trees, leaving little if any seed source for future generations. Dead crowns, even once the needles are shed, prevent full sunlight from penetrating to the forest floor. Surveys examining white spruce forests a decade after their death have documented little, if any, white spruce seedlings across thousands and thousands of hectares. In the meantime, these dead standing trees are kindling for wildfires, which are common in the Yukon.

Extensive spruce beetle outbreaks have been recorded throughout BC, Colorado, Wyoming, Idaho and Arizona. A further 1°C (1.8°F) increase in temperature is predicted to increase the probability of spruce bark beetle outbreaks and favour a one-year breeding cycle (doubling reproduction rates and causing an immediate surge in the number of emergent adults) in Alaska, the Yukon and high elevations in the lower-latitude forests. Sadly, by 2060 the range of Engelmann spruce, a principal host for spruce bark beetles, is projected to decrease by 47 per cent in the contiguous western United States.

Central and northern Europe have also faced

the onslaught of the European spruce bark beetle (*Ips typographus*), which in a warming world has killed at least 3 million hectares (7.4 million acres) of Norway spruce between 1990 and 2001. Global warming is increasing the frequency and intensity of storms, which are fuelling blow-down in forests across Western Europe, providing an ample food source for the European spruce beetles. In fact, one of the biggest storms in modern times occurred in southern Sweden in January 2005, levelling 75 million cubic metres (31.8 billion board feet) of Norway spruce. Those freshly killed trees, along with the longest, warmest summer on record in 2006, allowed regionalized beetle outbreaks to kill another 3.2 million cubic metres (1.4 billion board feet) of Norway spruce. In January 2007, yet another huge storm pummelled southern Sweden. The following spring came early and the combination facilitated more European spruce beetle outbreaks.

Of course, the changes in forests around the world are not limited to bark beetles. Warming winters across northern Asia have permitted the Asian gypsy moth (*Lymantria dispar*) – the most destructive defoliator of coniferous forests – an opportunity to expand its range into northern and northeastern Siberia, including the largest contiguous boreal forest on Earth, in the Kamchatka and Magadan region. In addition, the occurrence of fire has been increasing and the fire season is beginning significantly earlier than it did a decade ago. The warmer, drier climate, according to the Sukachev Institute of Forest (at Krasnoyarsk), is preventing tree seedling regeneration of burned-over areas on Siberia's southern forest edge, which are turning into grasslands.

In the far northwestern boreal forests of North America, the western spruce budworm (*Choristoneura fumiferana*) is also on a feeding frenzy. This aggressive defoliator is eating its way through Alaska, Oregon, Washington, Idaho, Montana, British Columbia and the Yukon. Recent warming in Alaska appears to have removed the cold environmental constraint that prevented outbreaks from occurring in the near Arctic. A predominant concern surrounding the boreal forests is that they contain 47 per cent of Earth's stored carbon. Rising temperatures in the Far North are favouring destructive bark beetles and defoliators, which are quickly turning carbon sinks into carbon sources. As these situations continue to intensify, they are fuelling heat-trapping greenhouse gases and Earth's temperatures are climbing even higher.

The Piñon Pines

The incredible piñon pines and their constant companions the junipers spread across the American Southwest, encompassing over 19 million hectares (47 million acres). Together, they make up the third largest wooded ecosystem in the USA. Importantly, the ecosystem revolves around the health and well-being of the piñons, as they nourish life throughout the Southwest.

Piñons and junipers have carved out an existence between the high deserts of the Southwest and the ponderosa pine forests of the Colorado Plateau. They live in the woodlands, where moisture is scarce and therefore tree crowns rarely touch one another or come close to forming an unbroken green canopy. Rather, clumps of trees that are 300 or 400 years old, with crowns like a parasol and thin trunks, grow very slowly. Open forest floors provide habitat for desert plants like yucca and prickly pear to live alongside the piñons and junipers. In the springtime, vibrant-red Indian paintbrush wildflowers adorn the piñon woodlands across the entire Southwest.

Pines date back about 180 million years to northern Asia during the Triassic Period of the Mesozoic Era. Not only did the pines provide shade for dinosaurs, they also helped modify the water and energy flow between the land and the atmosphere. Throughout the ages, climates changed and continents were pulled apart and thrust together. Pines migrated into North America across the Bering land bridge that connected Alaska with Siberia. They travelled as far south as Mexican peaks in the Rocky Mountain chain, where it is believed that the new species of piñon pines originated.

All pines – about 100 species worldwide – are conifers: cone-bearing evergreen trees with needles in a bundle (fascicle) that is attached to a shoot. There are usually two, three or five needles per fascicle. Only one species has just one needle per fascicle and it belongs to one of the 11 species of piñon pines. *Los piñónes* is the Spanish word for the edible seeds of desert conifers.

Single-leaf piñon pines live in the Great Basin and in the hills of the Mohave Desert, with scattered populations in Arizona and southwestern New Mexico. The other member of the piñon clan found across the American Southwest is the Colorado piñon. This two-needled piñon is familiar to many because of its commercial nut, the pine nut, which is more accurately called its seed. Colorado piñons reside in Colorado, New Mexico, Arizona and eastern Utah in sprawling woodlands on mesas and along the foothills of the southern Rocky Mountains.

Living side by side with the piñons are three species of junipers, infamously known for their berries that are the primary flavouring in the alcoholic beverage called gin (a shortening of the Dutch word for juniper, *jeneverbes*). Alligator junipers are the most abundant throughout the Southwest and widespread in Mexico, with northern ranges extending into north-central Arizona and New Mexico. They are found with both single leaf and Colorado piñon. The second species, one-seed juniper, stretches from central Colorado to central Arizona, throughout New Mexico and southward into Mexico. At low elevations, these trees eke out an existence in pure stands that are too dry for piñons to make a living. Utah juniper, the third species, is almost always found side by side with single-leaf piñon, making its home in the Great Basin. It too is found in pure stands at lower dry elevations in Arizona, California, Utah and Nevada.

Several species of woodrats, or packrats, inhabit the piñon woodlands, including desert woodrats, Stephen's woodrats and the white-throated, the Mexican and the bushy-tailed woodrats. Woodrats are prodigious hoarders, carrying twigs, shredded bark, cones, animal droppings, bottle caps, tin foil and a host of other delights back to their shelters, which are known as middens. Middens are communal dwellings with a designated latrine. Evaporated urine acts, inadvertently, to weld plant matter in the midden. Middens that were constructed in rock shelters or caves where they were protected from weather and erosion have provided scientists with a "looking glass back in time" to the end of the Pleistocene Epoch.

Fossil samples from middens have been radio-carbon dated to between 7,800 and 30,000 years ago. Woodrats have shown that piñon–juniper woodlands existed 545 metres (1,800 feet) above sea level in the New Water Mountains near Yuma, Arizona. Temperatures in the Sonoran Desert during that period were between 1.6 and 3.8°C (2.88 and 6.84°F) cooler than today, with 75 to 175 millimetres (3 to 7 inches) more rainfall. Today, the closest stand to the one from 7,800 years ago is found 90 kilometres (56 miles) north at an elevation of 1200 metres (3,937 feet).

The piñon–juniper woodlands play many invaluable roles. They provide essential shade, help build soils, prevent soil erosion and create microclimates by regulating water as it moves from soil to atmosphere. The woodlands also provide habitat and food (their seeds) to many critters, including piñon mice, bushy-eared Abert's squirrels, cliff chipmunks, rock squirrels, Uinta chipmunks, woodrats, black bears and desert bighorns. In addition to their protein-aceous seeds, piñons also offer a rich sugar food source in their living bark, or phloem, cells. Black bears and porcupines are major consumers of piñon phloem, along with – of course – the in-satiable bark beetles like the western bark beetles

(*D. brevicomis*), mountain pine beetles (*D. ponderosae*) and the fecund engraver beetles (*Ips* spp.).

Piñons pines co-evolved with an intelligent genus of birds known as corvids, which include piñon jays, scrub jays, Steller's jays and Clark's nutcrackers. Spend even an afternoon in the piñon–juniper woodlands and you will see flocks, sometimes in the hundreds, of piñon jays swooping adeptly a couple dozen metres above the woodland treetops. Raucous-voiced, gregarious and intrepid are apt descriptors of these important woodland birds that nest in piñons, junipers or higher-elevation ponderosa pines. They eat nutritious insects, berries and seeds, particularly piñon nuts. Research at the University of New Mexico discovered that the availability of the emerald-green piñon cones, with their sparkling beads of resin, actually stimulates testis development in male piñon jays at just the right time for late summer breeding.

Toward the latter part of August the piñon pines' wingless seeds, which take three growing seasons to develop, become ripe. The cones, however, remain shut tight. The flocks sense that some of the cones are ripe and begin to chisel into them with their strong beaks. Piñon pines produce two colours of seeds: light tan and chocolate brown. The light tan seeds are mostly empty and completely disregarded by the jays. The chocolate-brown seeds, on the other hand, are carefully examined for their weight first, and then a "click-test" is performed during which the bird rapidly opens and closes its mandibles. Presumably, the seed must make the right sound or it is discarded. The jays temporarily store 20 acceptable seeds in their elastic esophaguses. Seeds are then cached in the soil about 10 kilometres (6 miles) away in the nesting area. The birds remember where most, but not all, of the seeds are located.

This is one of nature's wonderful symbiotic relationships. The piñon pines provide large, nourishing seeds for the birds and spread cone opening over several months, which offers a long-lasting autumn food source that can easily overwinter when other foods are unavailable. In return, the piñon pines' wingless seeds, which cannot be disseminated by the wind or germinate on top of the soil due to the arid environment, are planted in moistened soil by the corvids. Sometimes mice and voles raid bird caches and other times the jays themselves forget some of the buried seed. The piñons rely on these occurrences to perpetuate themselves.

Piñons not only have relationships with woodland fauna, they also have a long history with humans. As far back as 13,000 years ago, across Colorado and New Mexico, the piñons sustained paleo-humans. Piñon charcoal and seed coats were found in fire pits of Gatecliff Shelter in central Nevada that date back more than 6,000 years. The Navajo and Zuni peoples used piñon for construction lumber, and the Pueblo and Navajo used the piñon pitch on their stone griddles because of its renowned non-stick properties, which incidentally inspired the modern product called Teflon. Piñon pines were also used by Native American peoples for communicating with the spirit world and were revered for their many medicinal properties, including use as a disinfectant for cuts and abrasions and for relief from headaches and lung congestion.

Piñon pines also provided an essential food source in harsh environments for the peoples of the Southwest. In early September, after harvesting the cones, native peoples used fire to melt the resin along the cone scale to extract the seeds. The seeds are edible raw, roasted or boiled, and often they were ground into flour. The Navajo also mashed the oil-rich seeds into a delectable butter. All 20 amino acids, the building blocks of proteins, are found in piñon seeds, nine of them in very high quantities. Piñon seeds are particularly high in tryptophan, which is known to help people fall asleep, as well as phosphorus, iron, vitamin A, thiamine, riboflavin and niacin.

Since the late 1800s, the piñon–juniper woodlands across the Southwest have been drastically modified. They have been used for livestock grazing and wood harvesting. Due to fire suppression, many of the woodlands are also far denser than they were prior to Euro-American settlement. Scientists know that the high density of trees in these woodlands increases tree stress, particularly competition for water, and weakens trees by exposing them to disease and insects. Fumes of sulphur dioxide from coal-fired power plants are further worsening the ill health of the overcrowded southwestern piñon–juniper woodlands.

Over the past 15 years, scientists around the globe, with the aid of satellite and aerial photography, have been observing some ominous patterns of change. Very recently, peer-reviewed scientific articles have begun to connect the dots and conclude that as temperatures rise, the Earth is experiencing a global-change type of drought. Higher temperatures alone can increase forest water stresses, independent of precipitation amounts. No forest on our planet is safe from the impacts of drought. Subcontinental droughts across the western US and southwestern BC have caused rapid tree deaths in all 76 old forests sampled. Droughts are drastically altering the water cycle, including the amount of snowfall that accumulates (snowpack). In addition, earlier spring melts and runoff are lengthening summer drought durations.

Nowhere in North America better exemplifies the effects of a global-change type of drought than the US Southwest. In 2002 the region experienced an extreme drought, and in many areas 90 per cent of the trees were killed. Water-starved, heat-stressed piñon were completely incapable of fending off the onslaught of *Ips* bark beetles, which during outbreaks can breed at least three generations in a year. In parts of New Mexico during the early 2000s, 90 per cent of the state's emblem tree, the piñon pine, were killed. Picturesque communities like the capital city, Santa Fe, were left surrounded by dead trees that became kindling awaiting a spark. Interestingly, Governor Bill Richardson conscripted prison inmates to remove dead trees encircling Santa Fe and cost-efficiently fireproof the community. This forward-thinking action of utilizing prison inmates has been proposed in BC and California to assist their communities that are vulnerable to the hazards of overcrowded, fire-suppressed forests at risk of wildfires.

Thanks again to tree rings, we know of three droughts that occurred in the US Southwest throughout history. One of the most severe ones in the last 500 years took place in the 1950s. Researchers discovered tree rings from ponderosa pines just above the New Mexico piñon–juniper woodlands which revealed that a brief but extreme drought, likely amplified by historical fire suppression, killed the ponderosa pines and in turn increased the density of the piñons and junipers. The drought triggered an infestation of voracious *Ips* spp. and *Dendroctonus* spp. bark beetles. In 1957 the drought broke and tree death immediately halted.

The drought from 1997 to 2006 in the Southwest, with its peak in 2002, was similar to the subcontinental drought of the 1950s, with the notable exception that it had significantly higher temperatures. The elevated temperatures were lethal to both the piñons and the junipers across the entire Southwest. Research at the University of Arizona's Biosphere 2 laboratory found that when temperatures increased by 4 degrees, piñon pines died five times faster. When the climate is warmer, it takes a much shorter drought to kill trees, which explains why the 1997 to 2006 drought in the Southwest – in tandem with bark beetles – killed millions more piñons and ponderosas than the 1950s drought did.

Ips and western pine beetles affected more than 3.2 million hectares (7.9 million acres) of piñon and ponderosa pines in Colorado, Arizona, New Mexico, Nevada and Utah between 1996 and 2005. Severe drought and increased temperatures contributed to widespread mortality. Across an astonishing 1.4 million hectares (3.5 million acres), 95 per cent of the piñons were killed. Water-starved pines cannot manufacture gooey pitch – their only defence against bark beetles – so they became sitting ducks for the inevitable ambush by the beetles.

Elevated temperatures have also enabled bark beetles that live in more southerly regions to march north into Arizona, southern Utah and Colorado. The round-headed pine beetles (*Dendroctonus adjunctus*) have attacked a number of pine species in the Southwest, including ponderosas. Populations of southern (*D. frontalis*) and Mexican (*D. mexicanus*) bark beetles have also erupted in Arizona. Unfortunately for the forests in the region, both the southern and the Mexican bark beetles have an extraordinary ability to produce in excess of five generations within one year.

Piñons and junipers each contend with drought in markedly different physiological ways. Piñon pines stop photosynthesizing altogether; that is, they disallow any moisture from leaving the tree. This strategy is initially effective, but ultimately the tree dies of carbon starvation. When water content in the soil drops below 15 per cent, piñons die. Junipers, on the other hand, continue to photosynthesize and allow precious water vapour to escape in a trade-off to acquire CO_2. They gamble that their roots will locate more water to replace the water vapour lost from the needles. Junipers are more drought-tolerant than piñons, but when the water vapour they release is not replaced, the xylem cells, which conduct water in columns, snap. Despite being more drought-tolerant than piñons, junipers too died en masse during the 2002 drought.

The extreme drought of 2002 has significantly altered the piñon–juniper woodlands of the Southwest. Not only did at least 100 million mature trees die, but invasive species like cheatgrass (*Bromus tectorum*) muscled into the ecosystem. Cheatgrass is highly flammable and thus provides the opportunity for more frequent fires on the landscape. Overall, woodland forests are dramatically changing, in some cases reverting to prairie. Post-piñon–juniper woodland ecosystems may soon, like those of interior valleys in California, resemble the wildfire-prone shrubland of the chaparral or the grassland ecosystem of the savannah.

Scientists from the University of California Santa Barbara and elsewhere examined almost 1,110 Douglas-fir tree rings from Arizona and mountain forests in New Mexico. Over the past 25 years, fire, drought and beetles have killed 18 per cent of the sites studied. The researchers predict that with rising temperatures and two more short but intense drought periods, more than 50 per cent of the remaining mountain forests in the Southwest could die. This type of devastation has enormous repercussions for all life, not only within the forest. Compromised ecosystem services include less control over snowpacks; an increase in downhill water discharge that removes the upper layers of soil; loss of biological diversity; a decrease in fresh oxygen being released into the atmosphere; and an increase in CO_2 as the forests release the gas instead of storing it. All of these factors feed back into one another and adversely affect the quality and sustainability of all life, including human.

Perhaps a more tangible example of how this changing world will alter the way we live is through water usage. Las Vegas, Phoenix, Tucson, San Diego and Los Angeles all depend on the forests of the Southwest to provide water for agriculture, electricity, industry and drinking. In 2008 the Scripps Institution of Oceanography at University of California San Diego predicted that, with rising temperatures and projected rates of water usage based on current consumption, there is a 50 per cent chance that the Hoover dam will not be generating power by 2017, and that Lake Mead, which feeds the Hoover dam, will run dry by 2021. People in the US Southwest are going to have to become far more efficient at how we consume water. As a matter of fact, across the US we use trillions of litres of water per day during the summer on urban landscape vegetation alone. That type of abusive practice will have to change ... quickly.

The Whitebark and Limber Pines

Two species of tenacious, five-needle, high-elevation treeline pine – one from the Old World, the other from the New – cling to the peaks along the northern Rockies and Cascade–Sierra system: limber and whitebark pines. Miraculously able to live for well over a thousand years, these trees create microclimates and act as phenomenal grocery stores that sustain local wildlife. Their life story is intriguing and their value to western North America is priceless. Yet their very existence is perilously close to its end, as rising temperatures have enabled their predator – the mountain pine beetle – an opportunity to wipe them off the face of the Earth.

Limber pines are believed to have evolved in the Mexican Highlands. As with all other long-lived organisms, climate change brought an opportunity to expand their range – with the assistance of jays at first – and they began to advance northward. Today, these species of pine exist at or near treeline in the Rocky Mountains and in intermountain ranges from southeastern BC and southwestern Alberta down through Oregon, Idaho, Montana, the Dakotas, Nebraska, Wyoming, Colorado, Utah and Nevada to northern New Mexico and through northern Arizona to southern California.

Whitebark pines, on the other hand, belong to a group of stone pines that span the Old World from western Europe to far-eastern Siberia and into Japan and Korea. Genetic markers elegantly show that whitebark pine diverged from the Eurasian stone pines between 1.3 million and 600,000 years ago. The Old World stone pine journey across the Bering Strait was made hand in hand with its faithful partner, a member of the *Corvidae*, or crow family, the nutcracker. The Eurasian nutcracker differentiated into a new species, called Clark's nutcracker, which have become the whitebark pines' North American partner. Whitebark pines have a larger range than limber pines, from the Bulkley Ranges in BC to the northeast Olympic Mountains and the

Cascade range in Washington and Oregon to Kern River of the Sierra Nevada, and from the northern Rockies in BC and Alberta into Idaho, Montana, Wyoming and Nevada.

Limber and whitebark pines are keystone species. That is, they play a crucial role in maintaining the structure of high-elevation ecological communities. In fact, Clark's nutcrackers and these pines build entire ecosystems and feed them. In order to understand these thrifty pines, we need to examine their cones and seeds. At least one theme is applicable across the game-board of biology: long-lived organisms are in no hurry to reproduce. Whitebarks, for instance, do not produce their first cones until they are about half a century old. In some cases, it takes them at least that long just to grow above the winter snowpack.

Both limber and whitebark pines take around 25 months to produce viable seeds in their cones. The stalkless cones are held upright on their branches. In late July or early August the purple cones glisten with droplets of terpene-laden pitch. Usually by the end of August or early September the large, wingless, fatty and protein-rich seeds are mature. Some of the tight cone scales open marginally, but not far enough for the seeds to dislodge and fall to the ground. The protein-loaded whitebark pine seeds contain 16 amino acids, with glutamic acid, lysine and arginine in high concentrations. They are an excellent food source for wildlife (and people too).

Both whitebark and limber pines co-evolved with corvids, as they depend upon the clever birds to disseminate their seeds. Clark's nutcracker is their primary avian partner, a species named on August 22, 1805, by explorer William Clark in Lemhi Pass of the Bitterroot Mountains, on the present-day boundary between Montana and Idaho. These unmistakable birds with their *kra-a-a* call are native to the western states and western Canada. Clark's nutcrackers are 30 centimetres (12 inches) long, with a wingspan exceeding 55 centimetres (22 inches). They weigh 140 grams (nearly 5 ounces) and they have a distinct bill that is stout, black and chisel-like and measures 31 millimetres (1.2 inches) from nostril to tip.

These birds splendidly evolved to harvest limber and whitebark cones with their powerful beaks that accurately jackhammer through sealed cone scales. They even have a specialized sac-like extension called a diverticulum, located on the floor of their mouth, under the tongue, which stores the large, wingless pine seeds. Like the piñon jays, Clark's perform both visual and auditory seed checks before pouching limber or whitebark seeds. These birds are highly selective foragers who aren't in the business of transporting or eating infertile or low-calorie pine seeds. When packed full with about 82 seeds, their elastic pouch stretches to the size of a walnut and weighs an additional 31 grams (1 ounce). Despite this impressive payload, Clark's are still able to attain a top speed of 50 kilometres per hour (31 miles per hour) and fly as far as 32 kilometres (nearly 20 miles).

During the summer Clark's augment their pine-seed diet with insects, spiders and even small mammals and amphibians. But as full-time residents of high mountain forests that are viciously frigid in winter, the birds are vitally dependent on pine seeds. Specifically, because they are very early breeders – laying between two and four brown-spotted pale-green eggs in late

February that hatch about 17 days later – their only available food source at that time of year are the highly nutritious pine seeds.

Clark's begin to harvest whitebark and limber cones in late August, working individually or in flocks of up to 150 birds. By mid-October they have accomplished their task. Sometimes they fly some 200 metres (660 feet) from the harvested trees, but other times much farther. They stop, look around, *kra-a-a a* few times to see if any Steller's jays or crows are watching them, and then carefully cache between one and four seeds, one at a time, into the earth about 25 millimetres (1 inch) deep. Then off they fly to do it all again until they empty their pouch completely. One Clark's nutcracker can easily place 98,000 whitebark or limber seeds in over 30,000 caches and later recover half for food for themselves and their young.

All corvids (ravens, crows, jays, magpies and rooks) have the largest cerebral hemispheres, relative to their body sizes, of all the world's birds. Clark's are no exception. These extraordinary birds are smart and possess exceptional long-term memory. Experiments have demonstrated that Clark's nutcrackers are able to remember where their caches are by triangulating the locations from triple visual cues like boulders, trees and mountaintops.

Clark's have direct competition in seed harvesting from red squirrels, which spend upward of 60 per cent of their foraging time on whitebark seeds. These critters are known to gather in excess of 800,000 seeds per animal and hoard their cones in large middens on the forest floor next to decaying logs, stumps or uprooted trees. A host of other animals also rely on the bountiful whitebark and limber pine seeds as food. Steller's jays, ravens, crows, mountain chickadees, red-breasted nuthatches and pine grosbeaks are amongst the other bird species that lay claim to these seeds. Chipmunks too are potential dispersers of whitebark and limber seeds, but only over short distances.

Observations and detailed studies examining bear scat have revealed that both black bears and grizzly bears, particularly the interior populations, travel to the high-elevation whitebark and limber forests in the autumn to raid squirrel middens and feast on pine seeds. When breeding, in order for bear sows to implant a fertilized egg successfully, which can be delayed for up to five months, they must have sufficient fat reserves to give birth while hibernating. Once born, the hungry cubs require rich mother's milk with fatty concentrations exceeding 30 per cent. A good supply of whitebark and limber pine seeds can actually determine the fate of bear populations, as without sufficient sustenance the sows will not implant and no new cubs will be born. These seeds are an imperative food source in Yellowstone National Park for interior grizzly populations. Besides raiding squirrel middens, grizzlies also harvest whitebark cones directly from the trees. It is truly breathtaking to watch these huge bears gingerly extract and eat the delectable seeds while deliberately avoiding the cone scales.

Clark's nutcrackers are directly responsible for creating all-aged whitebark and limber pine forests that are shade-intolerant plants as seedlings. As the pines grow, they eventually provide the shade that enables Engelmann spruce and subalpine fir to come up underneath. Water passes through the pines, spruces and firs, creating a microclimate of cooler summers, warmer winters and less wind. Whitebark and limber pines also build soils and provide shade for snowpack to retain its snow and release it slowly in the springtime. This in turn regulates the flow rate and water temperature of streams that sustain fish such as trout.

These forests are unique because of the gnarly high country where they grow. Life is tough, with poor soils, high-ultraviolet light, frigid temperatures, high winds and no thermal cover. But eventually, open patches coalesce and form continuous high-mountain forests. Then, one hot summer's afternoon, a bolt of lightning ignites the forest and resets its biological clock. Whitebark and limber pines, with their steadfast companions the Clark's nutcrackers, once again reforest the high-altitude, burned-over lands.

Prior to European settlement across the West, lightning-generated fires occurred in these mountain forests at frequencies ranging from once in 75 years to once in 200. Native North Americans would intentionally set fires to improve forage for game and create pasture for horses. These relatively more frequent surface fires, along with lightning fires, created a diverse landscape of forest species and age classes. In the high country, fires created openings for Clark's to cache whitebark and limber seeds. Some of those pine seeds germinate, requiring full sunlight to develop into saplings. Then, when about a half-century-old, they begin reproducing. These pines cannot regenerate under a shady canopy. Thus, fire suppression across western North America has changed the composition of many whitebark and limber pine forests, resulting in far fewer young pines and mostly older, mature pines.

Over the past century, whitebark and limber pine forests have changed considerably. Rising temperatures in the high country in the United States – during the past three decades particularly, though not exclusively – have unleashed insects in numbers never before seen. In many ways, these unique, high-elevation, slow-growing, long-lived trees are the 21st century canaries in the coalmine, signalling massive, potentially disastrous change on Earth.

Not all threats to the pines are recent. In 1910, for instance, a thousand French white pine seedlings were offloaded from a Romanian freighter at the port of Vancouver. The seedlings were infected with white pine blister rust (*Cronartium ribicola*), whose wind-spread spores quickly found a secondary host – the leaves of *Ribes* shrubs (gooseberries and currants). Alternating between its two hosts, the pines and the *Ribes,* this human-introduced pathogen took off like wildfire, killing all five-needle pines it came in contact with. By 1960 white pine blister rust had infected the range of western white pines in British Columbia, Washington, Oregon, Idaho and Montana. Sugar pines, the tallest of all pines in the world, were infected in northern California and the blister rust crossed the Continental Divide in Montana, infecting both limber and whitebark pines. Today, blister rust is ubiquitous throughout high-elevation limber and whitebark pines, killing all it comes in contact with. This introduced disease is yet another threat to the life of our once-pristine forests.

The most recent threat to the pines are the mountain pine beetles, which enter these forests and kill the pines at a rate that is truly astounding. If they continue on their present course, some scientists believe we are witnessing an irreversible ecosystem regime shift, impoverishing the mountains of western North America. Temperatures have risen by more than 2°C (3.6°F) in the mountains of the US, while in Canada the change is slightly less but the trend is certainly upward. Mountain pine beetles relish the opportunity to feed and breed on whitebark and limber pines. In fact, whitebark pitch is loaded with myrcene, which appears irresistible to the ravenous bark beetles.

Some scientists hypothesize that the whitebark pine was widespread throughout the West at one time, as evidenced by paleo-pollen samples taken from alpine lake cores. They believe that the mountain pine beetles survived the Pleistocene Epoch in some ice-free sanctuary. Then, as conditions warmed 12,000 years ago – and as in an epic battle scene, albeit perhaps a slow-moving one – the beetles ambushed the whitebarks and wiped them out over much of their lower range. The whitebarks' only refuge was in the high mountain forests, where they could survive but the beetles could not, apart from some years when rare milder temperatures permitted. With the recent increase in temperature, the protective cold barrier has lifted. The beetles have resumed the task they started thousands of years ago at lower elevations, eating and breeding in both the whitebark and limber pines.

In the US, mountain pine beetles have sped up their lifecycle to take full advantage of the removal of the cold barrier in the high mountain forests. At least a quarter of the Canadian mountain pine beetles have begun to do the same. In the past, mountain pine beetles were constrained to a two-year lifecycle whereby the adult beetle was forced to overwinter a second year before mating. These adults usually perished from frigid temperatures, as they are far less cold-tolerant than larvae. This constraint kept the beetle population in check. Today, rising temperatures have allowed the mountain pine beetles to breed and mature within one year. As temperatures in the northern high mountains climb, the US beetles' Canadian brethren are also predicted to begin reproducing within a single year.

Mortality rates in the whitebark and limber pines are as high as 90 per cent in the US. From surveys in northern BC, mortality rates of these pines range from 72 to 80 per cent. The mountain pine beetles are on a tear in the high country, similar to that experienced in the lodge-pole pines in the mid-2000s in BC. Death rates are doubling because the ecological regulation of cold temperatures has been lifted. For example, in 2007, the mountain pine beetles killed to 94,000 hectares (232,279 acres) of limber and whitebark in Wyoming. By 2010 over 450,000 hectares (1.1 million acres) had been massacred. The mountain pine beetles have now killed more whitebark and limber pines than all the trees burned in the epic 1988 Wyoming fires.

There is another distressing reverberation from the loss of whitebark pines for Wyoming and the nation's oldest national park – Yellowstone. The grizzly bear population relies heavily on the nutritious whitebark pine seeds in order to breed successfully. As mentioned earlier, without the rich fat content of the seeds, bear sows are often unable to consume enough nourishment for their bodies to enable uterine implantation after breeding. Furthermore, even without consider-ing reproduction, grizzlies need to eat enough high-calorie food to gain enough weight to last them through hibernation. Researchers know that three times as many grizzlies die in bad whitebark pine years as good ones.

The importance of every single whitebark pine cannot be overstated. In the tree's lifetime, say 700 years, it can provide over 504,000 individual seeds. Not only do those seeds create entire treeline ecosystems, but each tree goes on to act as a healthy restaurant that feeds a whole neighbourhood of critters. The well-being of the whole community relies on these pines.

Across much of western North America, these thrifty trees – sentinels of the high country – have become the tsunami sirens of global warming, showing ecologists, climatologists and physiologists that a warming world is irrevocably altering the landscape across an entire mountainous region of western North America. The death of these trees amounts to so much more than just the loss of two pine species. These forests provide ecosystem services that are crucial for collecting, holding and slowly releasing fresh water that sustains agriculture, communities and economies. Instead of excelling at their age-old tasks, millions of dead mountain pines are now becoming kindling – ready for future wildfire.

The Bristlecone Pines

I encourage everyone to find a special place in nature where, with just a little practice, you can feel, smell, hear and even taste the untamed universal energy that courses throughout our planet. For over the past two decades, with a short break to undertake my doctoral research in the snow gum forests of the Victorian Alps in southeastern Australia, I've made an annual pilgrimage to the White Mountains of east-central California. It is there that the oldest living single-stemmed trees – the Great Basin bristlecone pine – have been thriving for over 48 centuries. For me, these bristlecone pines offer a glimpse and insight into immortality. At elevations of 3500 metres (11,500 feet) above sea level, under the most extreme tree-growth conditions on the planet, these venerable vanguards are showing tree and climate scientists, very clearly, just how quickly the climate on Earth is changing.

When I carefully climb above 3500 metres in the White Mountains – with my protective eyewear, sunscreen, water, a snack and my notebook – I'm not only gazing at the ancient, weatherbeaten, russet-coloured, gnarled Great Basin bristlecone pines. I am also listening for buzzing wasps and solitary bees, the intense hum of cicadas, the *kra-a-a* of Clark's nutcrackers or the chirps of the golden ground squirrels. Have you ever listened to the song of the wind? I'm always looking for deer tracks, ground squirrels, lizards sunbathing, birds soaring. If I'm fortunate and very quiet, I'll catch a glimpse of a red-tailed hawk or a majestic golden eagle. The redolent smells of the Great Basin bristlecone pine are interspersed with wafts of sagebrush, Indian paint brush, pennyroyal, mountain mahogany and wild snapdragons. In the extreme environment – arid, freezing, often blasted by gale-force winds that can carry shards of ice or grains of sand – all of life works in concert to make a living within a limited window of time. To me, this is nature's symphony.

Of the hundred or so kinds of pines on the globe, three species are grouped into the foxtails: foxtail pines, Rocky Mountain bristlecone pines and Great Basin bristlecone pines. Pines that belong to the group of foxtails are so named because of their densely packed five needles wrapped in a fascicle sheath, resembling the bushy tail of a fox. All foxtails live on mountaintops in the southwestern United States, in distinct populations and regions. As an ecostress tree physiologist, my passion is to understand how life at the edge of treeline makes its living. I'm drawn to these trees like steel to a magnet.

The first species, aptly called foxtail pines, are found at elevations above 2000 metres (6,560 feet) along the Oregon–California Klamath range, in two major populations that are about 475 kilometres (295 miles) apart. The distinctive Californian Sierra Nevada populations have bright red-orange bark and make their living at between 1800 and 3400 metres (5,900 to 11,000 feet) above sea level. There are some magnificent specimens in Sequoia National Park that have been alive for over 2,100 years.

The second members of this elite group are the Rocky Mountain bristlecones. These pines are found living with and above another of the five-needle pines, the limber pines, but the prickles on the Rocky Mountain bristlecone easily distinguish it. In addition to prickly cone scales, the tasselled needles at the end of the branches are covered with copious amounts of pungent, dried-white resin on needle surfaces. Rocky Mountain bristlecone pines grow at elevations between 2270 and 3800 metres (7,450 and 12,470 feet) above sea level in south-central Colorado, just south of Rocky Mountain National Park, along Pikes Peak south to Spanish Peaks, on Mount Evans, and along the Sangre de Cristo Mountains in northern New Mexico. There is also one outlier population outside Flagstaff, Arizona, on the San Francisco Peaks. These Rocky Mountain bristlecone pines can thrive for well over 2,400 years, the equivalent of 876,000 sunrises.

Great Basin bristlecone pines, third in the foxtail group and the one I visit in the White Mountains, grow at elevations ranging from 2700 to 3500 metres (8,860 to 11,480 feet) above sea level in eastern California's White, Inyo and Panamint mountains. They also grow along many high mountain peaks in Nevada, including the White Pine Mountains, the Quinn Canyon, Fish Creek, Schell Creek and Snake ranges, the

Charleston Mountains (and elsewhere). In Utah, they reside along the Indian and San Francisco peaks, the Pine Valley and Wah Wah Mountains, the Deep Creek range and on the Markagunt and Aquarius plateaus. The Great Basin bristlecone pines are unmistakable trees. At lower elevations, they mostly mingle with limber pines and infrequently overlap with whitebark pines. Their cones are thinner and bristly, and when maturing they exhibit iridescent, purple-coloured cone scales with the most sublimely aromatic pitch I've ever encountered.

Fossils from Idaho, Nevada, Colorado and New Mexico show that foxtail pines have existed in the western mountains for at least the last 45 million years. These trees epitomize thrift, with some incredible adaptations to conserve energy and maintain growth despite the fierce odds. For instance, the growing season along the White Mountains is about 45 days per year, with a meagre 325 millimetres (13 inches) of precipitation, 80 per cent of which falls as snow. Great Basin bristlecones save energy by retaining needles for over 45 years. These needles are exceptionally well designed for the long haul and can sustain 160-kilometre-per-hour (99-mile-per-hour) winds carrying sand or ice particles.

In this arid environment, the needles require specialized water-saving features to conserve every last drop of moisture these trees can find and store. These unique needles have sunken CO_2 exchange mechanisms, known as stomates, with loose plugs of wax to safeguard the inner needle surface from water loss. They are also coated with thick outer layers of protective wax and are tightly packed. Foxtail-like configurations along the branches serve to protect them from extreme mountaintop winds. Another tremendous trait among the foxtail clan, displayed by no other pines, is the ability to grow a new bud that can develop into a new branch from any bundle of needles. Hence, a branch can perpetually replace itself. Furthermore, these pines can replace broken crowns by initiating new branches. These are resourceful mechanisms that ensure extreme longevity, with trees growing for thousands of years.

Temperatures in the warmest months of the year rarely exceed 10°C (50°F). The foxtails were built to develop slowly, except when it comes to manufacturing cones, which they produce regularly. Even 4,500-year-old trees – miraculously – make viable seeds. Producing cones costs the trees their valuable stored carbohydrates, but that is a risk they assume despite low temperatures. Pollen-shed and wind-borne pollination occur in mid-July to early August, which is late for a pine. Their winged seeds mature and cone scales open for wind dissemination in late September and early October of the next year. Some scientists have hypothesized that the deep purple colour of the developing cone allows them to absorb more heat from the sun, compared to green cones, speeding up development by perhaps as much as 25 per cent. For these pines in cold climates, their seeds require no chilling, meaning they are able to germinate immediately provided they find moisture and exposed mineral soil.

The foxtails also receive a helping hand from the industrious Clark's nutcrackers. I have spent countless hours in these mountain forests listening and watching the loquacious Clark's. They are drawn upward to these ancient forests from the piñon and limber pine forests at lower elevations because of the predictable and constant seed production of the Great Basin and Rocky Mountain bristlecone pines. Bristlecone pine seeds are barely one-twentieth the weight of piñon seeds, but still they contain the life-sustaining proteins, carbohydrates and fats needed for rearing early-spring chicks. Some scientists, including myself, believe that the Clark's nutcrackers were responsible for planting these ancient mountain forests and perpetuating them.

One of the most fascinating aspects of visiting the ancient Great Basin bristlecone pines is their roots. Their enormous, exposed, structural lateral roots look like the arms of giant squid as they spread as far as 16 metres (52.5 feet) outward from the main trunks. Clearly these very old, weather-beaten, squat trees need anchoring against ferocious winds on steep, rocky slopes. I've observed extensive root grafting in Great Basin bristlecone pines and, with the aid of my field microscope, I've counted over 2,200 years of growth rings in one structural root. Perhaps even more intriguing is the fact that some roots can mine the same area of soil for centuries with the assistance of a symbiotic fungus called mycorrhiza. The microscopic fungus meticulously extracts phosphorus, nitrogen and water from the soil and passes it to the tree's root. In return, the tree feeds its partner with carbohydrates manufactured in the needles and sent to the furthest tips of its roots to explore and exploit more soil.

Two spectacular visual features catch my breath each time I explore these forests. First, the exposed wood on the ancient trees shows a kaleidoscope of colours along the spiralled grain of the wood. The colours range from black and gray to reddish-brown and russet to a resplendent golden-yellow. The wonderful colours are courtesy of a species of wood-decay fungus that works very slowly. If you happen to be in the high country with these extraordinary beauties as the sun is setting, I guarantee you will witness one of nature's most spectacular arrays of colour.

The other striking feature, which first caught my attention a of couple decades ago when I first visited these pines, are the thin strips of living bark. First to record this phenomenon was Dr. Edmund Schulman in the early 1950s, who described them as "life lines." These ancient trees can live for hundreds and perhaps thousands of years with only 25 per cent or less functional, living bark after the remainder of the tree has died. Closer examination reveals that living roots adjoining the living bark feed just that sector of the living crown. Unique to the foxtails is sectored architecture. If a big root dies, the trunk sector formerly supplied with water dies. In turn, so does the adjoining portion of the crown. There is a large degree of autonomy within these trees – a very rare trait in the arboreal world.

It is absolutely incredible that these trees live so long. They only live in the toughest mountaintop environments, where cold temperatures, aridity and intense winds keep most organisms from getting a toehold. The soils are mostly rocky and if the occasional fire even occurs, there is little, if any, surface fuel to burn. The climate is so dry that fungi, which normally rot wood, can grow imperceptibly. These trees produce copious amounts of gooey pitch throughout their entire life as a protection against infrequent insect attacks. And, unlike any other known living thing, these trees and the ancient cliff cedar of the Niagara Escarpment show no signs of degenerative aging. The oldest trees live more than 3 kilometres (1.86 miles) above sea level and are bombarded by extreme cosmic radiation, yet they exhibit no mutations as they stoically enter their 48th century of life. Moreover, gerontologists are awed that there are no signs of any chromosomal changes, no shortening of chromosome tips as the trees age. Great Basin bristlecone pines eventually die only because they outgrow the very soil and rock that supports them.

Each year, tree species in the temperate and boreal forests grow a combination of early wood and late wood, forming one annual ring that normally encircles the tree just beneath its bark. Tree growth, like all plant growth, depends upon both environmental and physical factors, such as temperature, sunshine, wind, soil properties, steepness of slope, snowfall and rainfall. During the growing season, when snowmelt and rainfall are plentiful, growth rings are widely spaced. Conversely, when there is a drought, tree-ring growth is significantly diminished. Tree growth in stressful environments, such as rocky soils, along cliff faces or at the tops of mountains, is very sensitive to environmental factors and displays excellent variation in ring-to-ring growth – known as variation sensitivity. It is these tree rings that scientists study.

Interestingly, Leonardo da Vinci examined tree rings when he studied growth patterns in the 1400s. But it wasn't until 1894 that a University of Arizona astronomy professor, Andrew Ellicott Douglass, began to decipher how tree rings could be used to date structural beams from Native American ruins in the southwest United States and determine ages of the colossal Sierra Nevada sequoias. Douglass founded dendrochronology, the science of tree-ring patterns. Scientists use a non-destructive method of sampling whereby they carefully extract pencil-thin cores from living trees. By counting each annual ring, which contains a lighter band of early wood and a darker band of late wood, researchers are able to accurately determine the tree's age. Tree-core samples are then compared to form a chronology, an arrangement of events in time. By matching overlapping patterns of tree ring growth, dendrochronologists use a technique called cross-dating. It's this precise method that has enabled scientists to obtain a continuous look back in time from Great Basin bristlecone rings of the White Mountains to 8,800 years ago.

The oldest Great Basin bristlecone pines live within 150 metres (500 feet) of treeline, which is the highest elevation at which they can exist – about 3500 metres (11,500 feet) above sea level. These Great Basin bristlecones are telling dendrochronologists the extent and abruptness of global warming over the past 50 years. Two sites in northeast Nevada and one sampling area in eastern California examined 20,000 tree rings of Great Basin bristlecone pines at 3200 to 3600 metres (10,500 to 11,800 feet) above sea level. They found that unprecedentedly wide rings of tree growth have occurred since the mid-1950s, compared with the previous 3,500 years. This stark anomaly coincides with temperatures having risen by 2°C (3.6°F) in these mountainous regions of western North America. Over the past three and a half millennia, none of the tree-ring growth matches what has occurred over the past five and a half decades.

Global warming is extremely dangerous for these ultimate mountaintop specialists of our planet, as they only know how to make haste slowly. They have never had to contend with the likes of the mountain pine beetle, because for the past 11,000 years it has been far too inhospitable for beetles to ever venture up to the tops of these frigid, dry, windswept mountains. Distressingly, this appears to be changing. In Colorado, mountain pine beetles have killed over 110,000 hectares (272,000 acres) of the intimately intermingling limber and Rocky Mountain bristlecone pines, and it seems they have only just begun to ramp up their attacks. In 2005 small outbreaks of mountain pine beetles were reported in the Great Basin bristlecones in Nevada's Snake range and in Great Basin National Park. California's Great Basin bristlecones are being monitored very closely for any indication of these lethal beetles.

In the meantime, the foxtails have another potentially deadly attack to contend with. All three of the foxtail clan are susceptible to the introduced (rather than native) white pine blister rust mentioned earlier. Blister rust is fatal to the bristlecones too and has recently made forays into stands of them in the Sangre de Cristo range of New Mexico. Global warming opens up avenues for wood-decaying fungus to accelerate attacks.

In a warming world, bristlecone pines are in jeopardy. Already very close to their respective summits, they have no higher ground to conquer. They are already teetering on the edge. Rising temperatures appear to be speeding up bristlecone pine growth. Faster growth means that the pines risk using up their carbohydrate reserves, as these trees were built to grow slowly over thousands of years. Higher temperatures are often associated with drought and these high-mountaintop ecosystems are already dry, only just able to supply the necessary moisture for slow growth. Already since 1951, the number of days that snow sits on the ground in California's high mountaintops has decreased by 16. This means that soil moisture toward the end of the growing season will be in short supply, very likely placing the trees under water stress.

Great Basin bristlecone pines are finely tuned, cold-adapted organisms that, when faced with rising temperatures, run the risk of starting growth too soon in the spring or continuing growth too late in the fall. Early growth is also at risk of being hit with a late frost, which could impair reproductive buds and ultimately prevent bristlecone pines from producing viable seeds. These extraordinary trees are facing an uncertain future, which contradicts their role as the gate-keepers to the Holy Grail: the secret to eternal life.

pines a
eir respec-
round t

Our Future

One of the privileges of being a scientist in the United States is the opportunity to belong to the American Association for the Advancement of Science. Our weekly magazine provides a portal for world-class, peer-reviewed science and also insight into society at large. As a science communicator, I am an ambassador for my profession. Science has played a significant role by shaping every facet of our world as we know it today. As a profession, we are rigorously trained to be cautious and meticulous. We are curious by nature and our business is knowledge. Knowledge is power. And it empowers us to take action when necessary.

When my publisher, Don Gorman, first suggested that I undertake this bark beetle manifesto, I was thrilled at the opportunity to examine our forests of western North America. After a couple of months of reading a couple of thousand scientific papers and several dozen books, I was shocked at what was going on in our forests – not only in western North America, but worldwide.

Sometime later this year or early in 2012, the seven-billionth human will be born on Earth. In my opinion, humans are exceptional problem solvers – it's what we do best. Irrespective of what profession one is involved with, the most innovative information leads to the best decision.

Mature forests on every forested continent have begun to die. Each day, 82 million metric tons of heat-trapping CO_2 is being belched into our biosphere. Rising temperatures are creating droughts that are killing billions of mature trees in the Amazon jungle and elsewhere around the globe. Some 40 per cent of the oceans' phytoplankton has died from warming currents preventing upwelling and denying nutrients to the base of the entire marine ecosystem. Coal burning is releasing 3,000 metric tons of mercury vapour annually, and if left unchecked this amount is predicted to increase by at least 25 per cent by 2020. Mercury is an awful neurotoxin that has already left a deep footprint throughout the entire Arctic ecosystem and is evident elsewhere as it freely circulates through the oceans and insidiously in-filtrates fish. Over the past 150 years we have fished out our oceans, yet at least 2 billion people rely on these dwindling stocks as their only daily source of protein. Knowing this information – and that history has shown us that rising temperatures spur on droughts and droughts annihilate societies – leads us to the most important decision in the history of humankind. We must plan for future climate disruption.

Interesting to me, being a wordsmith, is that it's all in how we pose the most important problem. Special-interest groups, and their high-priced lobbyists in alligator shoes, have chosen to denigrate science and scientists in an effort to hoodwink the public into thinking global warm-ing is a political issue that, if dealt with, is apt to take away their jobs. For example, when a recent poll asked Americans, "What do you think is the most important problem facing the country today?" only 1.5 per cent of the respondents offered global warming or the environment as an answer. However, global warming and environment was the most frequent response – more than tenfold higher than before – when the question was rephrased as: "What do you think will be the most serious problem facing the world in the future if nothing is done to stop it?" When the term "global warming" was exchanged for "climate change," even self-identified Republicans were more likely to consider climate change a real phenomenon.

Whether the public or politicians understand it or yet recognize it, the climate on Earth has begun to change significantly. The most precious commodity is not a barrel of oil, but rather a barrel of fresh water. Unbridled destruction taking place in exploiting the oil sands of northern Alberta is a repugnant reminder of human greed and blatant disregard for just how valuable 4.3 per cent of the Earth's fresh water supply will be in a warming world. Worldwide, the climate has begun to disrupt agriculture, driving prices of all commodities to record highs. How will China feed 1.4 billion mouths if the drought in its west continues?

Rising temperatures of 2°C (3.6°F) across the western United States have begun to kill 76 mature forests. In the Southwest, a deadly drought in the early 2000s carbon-starved at least 100 million piñon and juniper trees, allowing billions of bark beetles an opportunity to speed up their lifecycles and feast on the helpless trees. Scientists know that, with higher temperatures, trees die much quicker during a drought. Those higher temperatures will cause irreversible ecosystem regime shifts where piñon and junipers can no longer make a living.

At least another billion mature trees have died in the US from rising temperatures, droughts and voracious bark beetles. In fact, US Interior Secretary Ken Salazar has referred to bark beetles as "the [hurricane] Katrina of the West, and their devastation might just be beginning." What we are beginning to witness in the whitebark and limber pines is an assault by bark beetles of biblical proportions. Some populations will survive, but many will not. The ecosystem services, including capturing the diminishing snowpack and slowly releasing the springtime moisture to recharge creeks and rivers that feed our irrigation systems, are priceless. The fact that the most exquisite trees on Earth, the Great Basin bristlecone pines, are in a perilous position due to rising temperatures and an emerging onslaught of bark beetles has kept me awake for far too many nights.

Rising temperatures, irrefutably stoked by

burning fossil fuels, have enabled bark beetles to turn the forests of Alaska, Arizona, Idaho, Colorado and Wyoming from healthy carbon sinks into decaying graveyards that emit rising greenhouse gases. British Columbia has experienced the largest bark beetle feeding frenzy ever recorded in modern times and is also now an enormous source of heat-trapping CO_2. Moving farther northward to Alaska and the Yukon, spruce beetles have taken full advantage of rising temperatures by accelerating their lifecycles to feed and breed. Alaska's massive boreal forests are becoming another source of CO_2. Mountain pine beetles have successfully, for the first time in recorded history and likely ever, jumped into the jack pines of the Far North, beginning their march across the Canadian boreal forest to Labrador. In addition, there is a good chance they will infest the jack, red and eastern white pines of the lake states in the US northeast. Synchronous regional bark beetle outbreaks are indeed harbingers of global warming – climate change that is well underway.

The National Center for Atmospheric Research has predicted that the western two-thirds of the US will experience extreme droughts in the coming decades. This is a major concern for some 4 million homeowners across western North America whose properties back onto the wildland–urban interface, because drought begets wildfires. Wildfires release a type of smoke particle, an organic carbon aerosol, which researchers predict will increase by 40 per cent over the next 50 years as forest fires burn more frequently under drier conditions. More smoke and organic carbon aerosols will deleteriously affect the health of millions of people across the West, particularly those suffering from lung conditions such as asthma and chronic bronchitis, not to mention heart disease.

Many corporations, including Google, Intel, Apple, Disney, 3M, DuPont, Coca Cola, Whole Foods, Wal-Mart and Subaru among others, have embraced what Bill Coors, the grandson of Adolph Coors of Coors Brewing Company, realized in 1950: "All pollution and waste are lost profit." Coors observed that industrial companies were taking raw materials and fuels from nature, cycling products through the economy and then generating tons of garbage. In turn, the garbage was polluting the ground water. An "open loop" system exploits nature's resources and deposits toxic waste at both ends. With almost 7 billion people on our planet, we cannot do this any longer. A "closed loop" economy, on the other hand, is one where the full array of actual costs is accounted for within the system. Companies and consumers are rewarded for reducing waste and the environment is safeguarded. As Earth Day founder Gaylord Nelson famously put it, "The entire economy is a wholly owned subsidiary of the environment, not the other way around."

Nature has a warehouse of proven principles, as well as a research and development laboratory with five billion years of product development. Many corporations draw their ideas, information and inspiration from ecosystems like prairies, coral reefs and ancient high-elevation forests. Business, like nature, is a living system – creative, productive and resilient. All waste is lost profit, and all value is created by design and adaptation. The ability to learn is crucial for survival.

Knowing what is happening in our wild ecosystems empowers us to embrace change and make an effort to adapt in our own lives. Change is opportunity, or an opportunity in disguise.

There is hope if each of us lends a helping hand. In many cases, the changes necessary can be as easy as altering little things we do each day.

Another privilege of being a scientist is that I set aside time to visit with primary and secondary students, as well as my college students. A decade ago, North American society embraced the slogan "go big or go home" and many lived beyond their means. Today, the slogan is "less is more," and our youth firmly grasp this concept. Many children are very aware of what a carbon footprint is and they know how to calculate it. They understand that carbon-based energy is a cost to our environment and they are very aware of careful management of our natural resources and of the underlying principle of conservation. High school seniors selected colleges based on how green the campuses rate. Our youth clearly recognize that innovation is our best friend in the 21st century and that the bridge to tomorrow hinges on the pathway of efficiency. In many cases, efficiency is just altering our habits.

Children always ask me why oil, gas and coal companies are subsidized with billions of dollars when these companies collectively make hundreds of billions. Again, I'm heartened that our youth see that the real costs of energy are expensive, yet by subsidizing old, carbon-based polluters we are encouraged to waste energy instead of conserve it. Moreover, our youth are aware that if the real cost of energy were accounted for up front in the marketplace, then clean, green alternative energies could become affordable. Through innovation, we could create industries which would employ people and protect the environment. Children also realize that global warming is a citizens' issue and that we all need to lend a helping hand.

Professor Steven Chu, the US Energy Secretary and a Nobel Prize-winning physicist, was quoted in *Newsweek* as saying, "Science has unambiguously shown that we're altering the destiny of our planet." He is very concerned that the Earth is losing ice and that 94 per cent of ice older than five years that existed just three decades ago in the Arctic is now gone. Ice at the poles, in land glaciers and on mountaintops reflects incoming solar radiation and helps keep our planet from overheating. Professor Chu is a proponent of mimicking the missing ice on Earth by making all roofs and pavement white (or at least light coloured) to help reduce global warming by both conserving energy and reflecting the sunlight back into space. He calculated that this would be the equivalent of taking all the cars in the world off the road for 11 years. And just think of the jobs that would be created from such an undertaking, especially given high unemployment rates worldwide.

In July 2010, 300 scientists from 48 countries contributed to the annual State of the Climate report, published as a special supplement to the Bulletin of the American Meteorological Society. The report said that, of ten indicators that are "clearly and directly related to surface temperatures, all tell the same story: global warming is undeniable." For the first time in the history of our species, we are deliberately altering the destiny of humankind and life as we know it on the planet. As an intelligent problem solver, father, concerned citizen and Earth doctor, I choose to do the right thing. I trust that if you've come this far and followed my trip through our magnificent forests now being ravaged by bark beetles, you too will make the necessary changes, lend a helping hand and respect and protect our planet – the only home we have.

litres (
y in a home
es. That'

What can you do to make a difference?

Lots!

First things first. Once you have read this list, ask two people to ask two more people ... and so on and so forth ... to undertake some of the changes listed below. Together, we can help to reverse the tide, feel as though we are a part of the solution and truly make a difference!

* Calculate how much energy you use at home, travelling and at work, which is your carbon footprint (carbonfund.org). Once you determine how much you and your family are spending, create a strategic plan to begin to cut back on your energy usage. Simple changes can make a vast difference in your carbon footprint.
* Reduce what you use. Reducing is the most important way that we can all easily change.

For instance, buying quality products helps to reduce waste that we are putting into landfills. Quality products may cost more at first, but they last longer and save you money in the longer term (not having to buy inferior products over and over again). Think of other ways you can reduce your consumption and waste.

✳ Reuse what you can. Every year, North Americans drink more than 100 billion cups of coffee. Approximately 14.4 billion disposable paper cups are thrown away – enough cups, when placed end to end, to wrap around the Earth 55 times. Instead, get yourself a stainless steel travel mug. Many coffee vendors offer a discount, so within a year even the fanciest travel mug will pay for itself! Or you can take the proceeds from your coffee savings and buy organic cotton bags. Reuse a cotton bag instead of using the disposable plastic, single-use supermarket bags. Make it a habit to return them to the trunk of your car or store them near the doorway after unpacking your groceries so you never find yourself shopping without them.

* Toilets consume an average of 80 litres (21 gallons) of water per person per day in a home with no water-conserving fixtures. That's almost 30 per cent of the average home's per-person indoor water use. Consider installing low-flow toilets and showerheads and reduce one person's annual water use from 109,000 litres to 50,000 litres (29,000 gallons to 13,000 gallons). If installing new fixtures is not in your budget right now, find other ways to conserve, like filling a plastic bottle with sand or rocks and putting it in the tank of your toilet to reduce water usage. Or how about purchasing a sand timer (less than $5) that will help you and your family limit your shower time, saving you electricity and water. You'll notice an immediate saving on your monthly bill.

* Turn off the tap when you brush your teeth, and only run the dishwasher on the economy setting when it's completely full, again saving you money on your electricity and water bills.

* Check that your vehicles have their tires inflated to the manufacturer's suggested pressure. In doing so, you increase your kilometres

per litre by at least 4 per cent, saving up to 7 cents per litre (almost 28 cents per gallon) on gasoline. Also, make sure your trunk is kept empty (aside from your lightweight cotton shopping bags, of course), as extra weight reduces fuel efficiency.

- Ride a bicycle, walk or take public transit. Forty per cent of all car trips in North America are less than 3 kilometres (1.86 miles). Get exercise, reduce your greenhouse-gas emissions and save money on fuel.
- The average house emits about twice as much CO_2 as the average car. An energy audit will save you as much as 30 per cent on your yearly bill. Most utility suppliers offer a free walk-through to help you save a bundle of money.
- Roughly half of our homes' energy expenses come from heating and cooling, so make sure to keep your furnaces and air conditioners in their most efficient condition by having them serviced biennially and changing air filters at least twice a year.
- By setting your thermostat to 19°C (66°F) for winter and 26°C (79°F) for summer, you'll significantly reduce your costs.

- Wash your clothes in cold water only and hang them out to air-dry, at least in summer. You'll save more money and protect our environment.
- Use a "smart" power bar to plug in all your electronic devices that have an always-on, standby mode. That way, all your devices can be truly turned off, reducing your power bill by a further 5 to 15 per cent. "Phantom" electricity drawn by devices in standby mode wastes $5-billion worth of power per year across North America.
- Remember to turn off all lights when you leave a room, shut down computers and printers when not in use and unplug all cellular phones, laptops, cameras, MP3 players and toothbrush adapters.
- Compost all organic waste, and always recycle paper, cardboard, cans and bottles. This helps reduce the greenhouse gas emissions associated with landfills.
- Help our beleaguered honey, bumble and solitary bees by not using any insecticides, herbicides, miticides or fungicides in your yard. In addition, plant yellow and blue

flowers in large blocks, so as to provide a safe source of nectar and pollen for our bees.

- Plant a tree for every member of your family. Trees reduce heating and cooling costs around homes and buildings by as much as 40 per cent. They also draw CO_2 from the atmosphere, filter stormwater runoff, purify the air and provide habitat for many urban critters.
- Stay informed! Look for green community businesses and associations and go talk with them to find out what else you can do. Find ways to stay in the loop and keep the idea of living a green life on your mind. For instance, on Facebook, find and "like" pages or groups that deal with green living or sustainability and you will receive constant updates of what others are doing and reminders of what role you can play.

Cast of Plants
in alphabetical order by family

Betulaceae (Birch) Family
* Green alder: *Alnus viridis*

Cupressaceae (Cypress) Family
JUNIPERUS SPP.
* Alligator juniper: *Juniperus deppeana*
* One-seed juniper: *Juniperus monosperma*
* Utah juniper: *Juniperus osteosperma*
SEQUOIOIDEAE
(SUBFAMILY OF *CUPRESSACEAE*)
* Giant sequoia or Sierra redwood:
  Sequoiadendron giganteum
* Coast redwood: *Sequoia sempervirens*
OTHER *CUPRESSACEAE*
* Northern whitecedar or cliff cedar:
  Thuja occidentalis
* Western redcedar: *Thuja plicata*

Fabaceae (Legume) Family
🐜 Mountain acacia: *Brachystegia glaucescens*

Fagaceae (Beech) Family
🐜 Oak: *Quercus* spp.

Myrtaceae (Myrtle) Family
🐜 Snow gum: *Eucalyptus pauciflora*

Pinaceae (Pine) Family
LARIX
🐜 Siberian larch: *Larix sibirica*
🐜 Western larch: *Larix occidentalis*
PICEA
🐜 Black spruce: *Picea mariana*
🐜 Colorado blue spruce: *Picea pungens*
🐜 Lutz spruce: *Picea X lutzii*
🐜 Norway spruce: *Picea abies*
🐜 Engelmann spruce: *Picea engelmannii*
🐜 Sitka spruce: *Picea sitchensis*
🐜 White spruce: *Picea glauca*
PINUS
🐜 Eastern white pine: *Pinus strobus*
🐜 Foxtail pine: *Pinus balfouriana*
🐜 Great Basin bristlecone pine: *Pinus longaeva*

- Jack pine: *Pinus banksiana*
- Jeffrey pine: *Pinus jeffreyi*
- Limber pine: *Pinus flexilis*
- Lodgepole, or shore, pine: *Pinus contorta*
- Pitch pine: *Pinus rigida*
- Ponderosa pine: *Pinus ponderosa*
- Red pine: *Pinus resinosa*
- Rocky Mountain bristlecone pine: *Pinus aristata*
- Sugar pine: *Pinus lambertiana*
- Western white pine: *Pinus monticola*
- Whitebark pine: *Pinus albicaulis*

PINUS SPP.

- Single-leaf piñon: *Pinus monophylla*
- Colorado piñon: *Pinus edulis*

PSEUDOTSUGA

- Coast Douglas-fir: *Pseudotsuga menziesii*
- Bigcone Douglas-fir: *Pseudotsuga macrocarpa*

OTHER *PINACEAE*

- Atlas cedar: *Cedrus atlantica*
- True fir: *Abies*
- Subalpine, or Rocky Mountain, fir: *Abies lasiocarpa*
- Western hemlock: *Tsuga heterophylla*

Salicaceae (Willow) Family

POPULUS

- ❀ Balsam poplar: *Populus balsamifera*
- ❀ European aspen: *Populus tremula*
- ❀ Quaking aspen: *Populus tremuloides*

Bookshelf

Adams, Henry D., et al. "Temperature sensitivity of drought-induced tree mortality portends increased regional die-off under global-change-type drought." *Proceedings of the National Academy of Sciences* 106, no. 17 (Apr. 28, 2009): 7063–7066. Full text available at www.pnas.org/content/106/17/7063.full (accessed 2011-08-06).

Allen, Craig D., et al. "A global overview of drought and heat-induced tree mortality reveals emerging climate change risks for forests." *Forest Ecology & Management* 259, no. 4 (2010): 660–684. Abstract and full-text PDF available at http://is.gd/zNZJdG (accessed 2011-08-06).

Anderegg, William R.L., et al. "Expert credibility in climate change." *Proceedings of National Academy of Sciences* 107, no. 47 (Jul. 2010): E176. Abstract and full text available at http://is.gd/MkB7wr (accessed 2011-08-06).

Bark Beetle Outbreaks in Western North America: Causes and Consequences. Proceedings of a Bark Beetle Symposium held Nov. 15–17, 2005, at Snowbird, Utah (44 pp.). Edited by Hannah Nordhaus. Salt Lake City: University of Utah Press, 2009.

Barnett, Tim, and David Pearce. "When will Lake Mead go dry?" *Water Resources Research* 44 (Mar. 29, 2008): W03201. Full-text PDF available at www.colorado.edu/geography/geomorph/geog_5241.../barnett_08.pdf (accessed 2011-08-06).

Berg, Edward E., et al. "Spruce beetle outbreaks on the Kenai Peninsula, Alaska, and Kluane National Park & Reserve, Yukon Territory: Relationship to summer temperatures and regional differences in disturbance regimes." *Forest Ecology & Management* 227 (2006): 219–232. Full text available at www4.nau.edu/direnet/publications/publications_b/files/Berg-2006-Spruce.pdf (accessed 2011-07-14).

Borden, John H., et al. "Synergistic blends of monoterpenes for aggregation pheromones of the mountain pine beetle (*Coleoptera: Curculionidae*)." *Journal of Economic Entomology* 101, no. 4 (2008): 1266–1275. Abstract available at www.ncbi.nlm.nih.gov/pubmed/18767736 (accessed 2011-08-06).

Breshears, David D., et al. "Regional vegetation die-off in response to global-change-type drought." *Proceedings of the National Academy of Sciences* 102, no. 42 (2005): 15144–15148. Abstract and full-text PDF available at www.pnas.org/content/102/42/15144.abstract (accessed 2011-08-06).

———. "Tree die-off in response to global-change-type drought: Mortality insights from a decade of plant water-potential measurements." *Frontiers in Ecology & the Environment* 7 (2009): 185–189. Abstract and full-text PDF available at www.esajournals.org/doi/abs/10.1890/080016 (accessed 2011-08-06).

Broder, John M. "Past Decade Warmest on Record, NASA Data Shows." *The New York Times*, Jan. 21, 2010. Full text available at www.nytimes.com/2010/01/22/science/earth/22warming.html (accessed 2011-07-02).

Dai, Aiguo. "Drought under global warming: A review." *Climate Change* 2, no. 1 (Jan./Feb. 2011): 45–65. Abstract and full text available at http://onlinelibrary.wiley.com/doi/10.1002/wcc.81/full (accessed 2011-08-06).

Dunn, David, and James P. Crutchfield. "Insects, trees, and climate: The bioacoustic ecology of deforestation and entomogenic climate change." Sante Fe Institute Working Paper 06-12-055 (Dec. 14, 2006). Full-text PDF available at http://csc.ucdavis.edu/~chaos/papers/itc.pdf (accessed 2011-07-15).

Garbutt, Rod. "Yukon Forest Health Report 2003." Whitehorse: Yukon Dept. of Energy, Mines & Resources, Forest Management Branch (Mar. 2004). Full-text PDF available at www.emr.gov.yk.ca/forestry/pdf/forest_health_report_03.pdf (accessed 2011-07-18).

Goldenberg, Suzanne. "Greenland ice sheet faces 'tipping point in 10 years'." *The Guardian* (UK), Aug. 10, 2010. www.guardian.co.uk/environment/2010/aug/10/greenland-ice-sheet-tipping-point (accessed 2011-07-06).

Guha-Sapir, Debby, et al. "Annual Disaster Statistical Review 2010: The Numbers and Trends." Brussels: Center for Research on the Epidemiology of Disasters, Université catholique de Louvain (2011). Full-text PDF available at http://cred.be/sites/default/files/ADSR_2010.pdf (accessed 2011-08-06).

Halter, Reese. *The Incomparable Honeybee & The Economics of Pollination*. 2nd ed. Calgary: Rocky Mountain Books, 2011.

Hansen, James. *Storms of My Grandchildren: The Truth about the Coming Climate Catastrophe and Our Last Chance To Save Humanity*. New York: Bloomsbury, 2009.

Harmon, Mark, et al. "Effects on carbon storage of conversion of old growth forests to young forests." *Science* 247, no. 4943 (Feb. 1990): 699–702. Full-text scanned PDF available at academic.evergreen.edu/curricular/ftts/downloadsw/harmonetal1990.pdf (accessed 2011-08-06).

Hu, Feng Sheng, et al. "Tundra burning in Alaska: Linkages to climate change and sea ice retreat." *Journal of Geophysical Research* 115, G04002 (Oct. 5, 2010). Full-text PDF available at www.life.illinois.edu/hu/publications/Hu_et_al_2010.pdf (accessed 2011-08-06).

Jasny, Barbara R. "Framing the climate debate." *Science* 332, no. 6026 (Apr. 8, 2011): 151.

Johnson, Russ, and Anne Johnson. *The Ancient Bristlecone Pine Forests: Living Then, Living Now*. Aspects of research by C.W. Ferguson. Illust. and map by Jack Moffett. All photos not credited are by Russ and Anne Johnson. Bishop, Calif.: printed and distributed by Chalfant Press, 1970.

Kelly, Peter, and Douglas Larson. *The Last Stand: A Journey through the Ancient Cliff-Face Forest of the Niagara Escarpment*. Toronto: Natural Heritage Books, 2007.

Kilian, Crawford. "Climate and cholera." *H5N1* (blog), Feb. 17, 2011. http://crofsblogs.typepad.com/h5n1/2011/02/

climate-and-cholera.html (accessed 2011-07-08). *See also below at* Shah, Sonia.

Lanner, Ronald M. *Conifers of California*. 2nd ed. Los Olivos, Calif.: Cachuma Press, 2007.

————. *Made for Each Other: A Symbiosis of Birds and Pines*. New York: Oxford University Press, 1996.

————. *The Piñon Pine: A Natural and Cultural History*. Reno: University of Nevada Press, 1981. Preview text available at http://is.gd/GEpram (accessed 2011-07-04).

Ligon, J. David. "Reproductive interdependence of piñon jays and piñon pines." *Ecological Monographs* 48, no. 2 (Apr. 1978): 111–126. Abstract available at www.esajournals.org/doi/abs/10.2307/2937295 (accessed 2011-08-06).

Logan, Jesse A., et al. "Whitebark pine vulnerability to climate-driven mountain pine beetle disturbance in the Greater Yellowstone Ecosystem." *Ecological Applications* 20, no. 4: 895–902. Abstract and full-text PDF available at www.esajournals.org/doi/full/10.1890/09-0655.1 (accessed 2011-08-06).

Marchand, Peter J. *Life in the Cold: An Introduction to Winter Ecology*. 3rd ed. Hanover, NH: University Press of New England, 1996.

Maser, Chris. *Forest Primeval: The Natural History of an Ancient Forest*. Corvallis: Oregon Sate University Press, 2001. First published 1989 by Sierra Club Books.

McDowell, Robin. "Indonesia's last glacier will melt within years." Associated Press in *USA Today*, Jul. 1, 2010. www.usatoday.com/tech/science/

environment/2010-07-01-indonesia-glacier_N.htm
(accessed 2011-07-08).

Milne, Courtney. *The Sacred Earth*. Toronto: Penguin, 1998.

Mooney, Harold, et al. "Biodiversity, climate change, and
ecosystem services." *Current Opinion in Environmental
Sustainability* 1, no. 1 (Oct. 2009): 46–54. Abstract
available at www.sciencedirect.com/science/article/pii/
S1877343509000086 (accessed 2011-08-06).

Nikiforuk, Andrew. *Tar Sands: Dirty Oil and the Future of a
Continent*. Vancouver: Greystone Books, 2008.

Perry, David A., et al. *Forest Ecosystems*. 2nd ed. Baltimore:
Johns Hopkins University Press, 2008.

Romm, Joe. "NSIDC bombshell: Thawing permafrost
feedback will turn Arctic from carbon sink to source in
the 2020s, releasing 100 billion tons of carbon by 2100."
Climate Progress (blog) at *Think Progress*, Feb. 17, 2011.
http://is.gd/a5lTDm (accessed 2011-07-19).

Rosenthal, Elisabeth. "Heat damages Colombia coffee, raising
prices." *The New York Times*, Mar. 9, 2011. www.nytimes.
com/2011/03/10/science/earth/10coffee.html (accessed
2011-07-15).

Safranyik, Les, and Allan L. Carroll. "The biology and
epidemiology of the mountain pine beetle in lodgepole
pine forests." In *The Mountain Pine Beetle: A Synthesis
of Biology, Management, and Impacts on Lodgepole Pine*,
edited by L. Safranyik and W.R. Wilson, 3–66. Victoria,
BC: Natural Resources Canada, Canadian Forest Service,
Pacific Forestry Centre, 2006. Abstract and full-text PDF

available at http://cfs.nrcan.gc.ca/publications/?id=26039 (accessed 2011-08-06).

Safranyik, L., et al. "Potential for range expansion of mountain pine beetle into the boreal forest of North America." *The Canadian Entomologist* 142, no. 5 (2010): 415–442. Full-text PDF available at www.sysecol2.ethz.ch/Refs/EntClim/S/Sa146.pdf (accessed 2011-07-13).

Sands, Roger. *Forestry in a Global Context*. Cambridge, Mass.: CABI Publishing, 2005.

Schaefer, Kevin. "The Sometimes Frost: Interview with Kevin Schaefer," by Bruce Gellerman, aired the week of Feb. 25, 2011, on *Living on Earth*, a syndicated weekly radio program produced in Somerville, Mass., by World Media Foundation. Full-text transcript plus streaming and MP3 audio available at http://is.gd/NZHZho (accessed 2011-07-14).

Shah, Sonia. "Climate's strong fingerprint in global cholera outbreaks." *Yale Environment 360*, Feb. 17, 2011. http://is.gd/5N9aSF (accessed 2011-08-06). *See also above at* Kilian, Crawford.

Spracklen, D.V., et al. "Impacts of climate change from 2000 to 2050 on wildfire activity and carbonaceous aerosol concentrations in the western United States." *Journal of Geophysical Research* 114, D20301 (Oct. 20, 2009). Full-text PDF available at ulmo.ucmerced.edu/pdffiles/09JGR_Spracklenetal.pdf (accessed 2011-08-06).

Tripati, Aradhna K., et al. "Coupling of CO_2 and ice sheet stability over major climate transitions of the last 20 million years." *Science* 326, no. 5958 (Dec. 2009):

1394–1397. Abstract available at www.sciencemag.org/content/326/5958/1394.short (accessed 2011-08-06). *See also below at* Wolpert, Stewart.

Turetsky, Merritt R., et al. "Recent acceleration of biomass burning and carbon losses in Alaskan forests and peatlands." *Nature Geoscience* 4, no. 1 (Jan. 2011): 27–31. Full text PDF available at www.nature.com/ngeo/journal/v4/n1/full/ngeo1027.html (accessed 2011-08-06).

van Mantgem, Phillip J., et al. "Widespread increase of tree mortality rates in the western United States." *Science* 323, no. 5913 (Jan. 2009): 521–524. Abstract available at www.sciencemag.org/content/323/5913/521.abstract (accessed 2011-08-06).

Vander Wall, Stephen B., and Russell P. Balda. "Ecology and evolution of food storage behavior in conifer-seed-caching corvids." *Zeitschrift für Tierpsychologie* 56, no. 3 (Jan.–Dec. 1981): 217–242. Abstract (und Zusammenfassung) available at http://is.gd/NczlJe (accessed 2011-08-06).

Williams, A. Park, et al. "Forest responses to increasing aridity and warmth in the southwestern United States." *Proceedings of the National Academy of Sciences* 107, no. 50 (Dec. 14, 2010): 21289–21294. Abstract and full text available at www.pnas.org/content/107/50/21289.full (accessed 2011-08-06).

Wolpert, Stewart. "UCLA researcher finds CO_2 at highest levels in 15 million years." UCLA Institute of the Environment & Sustainability Newsroom, Oct. 8, 2009. www.environment.ucla.edu/news/article.asp?parentid=4676 (accessed 2011-07-10). *See also above at* Tripati, Aradhna K., et al.

About the Author

Dr. Reese Halter is an Earth doctor, an award-winning science communicator and voice for ecology, a distinguished conservation biologist at California Lutheran University and a broadcaster. He is a sought-after public speaker and founder of the International Conservation Institute – Global Forest Science, through which he regularly visits schools and encourages children worldwide to embrace conservation, science exploration and learning. Dr. Reese and his family live in Malibu, California. He can be contacted through www.DrReese.com.

Other Titles in this Series

The Incomparable Honeybee
and the Economics of Pollination
Revised & Updated
Dr. Reese Halter
ISBN 978-1-926855-64-6

Ethical Water
Learning To Value What Matters Most
Robert William Sandford
& Merrell-Ann S. Phare
ISBN 978-1-926855-70-7

Becoming Water
Glaciers in a Warming World
Mike Demuth
ISBN 978-1-926855-72-1

The Beaver Manifesto
Glynnis Hood
ISBN 978-1-926855-58-5

The Grizzly Manifesto
In Defence of the Great Bear
by Jeff Gailus
ISBN 978-1-897522-83-7

Denying the Source
The Crisis of First Nations Water Rights
Merrell-Ann S. Phare
ISBN 978-1-897522-61-5

The Weekender Effect
Hyperdevelopment in Mountain Towns
Robert William Sandford
ISBN 978-1-897522-10-3

RMB saved the following resources by printing the pages of this book on chlorine-free paper made with 100% post-consumer waste:

Trees · 10, fully grown

Water · 4,751 gallons

Energy · 4 million BTUs

Solid Waste · 302 pounds

Greenhouse Gases · 1,053 pounds

Calculations based on research by Environmental Defense and the Paper Task Force. Manufactured at Friesens Corporation.